CITIZEN GROUPS AND THE NUCLEAR POWER CONTROVERSY

The MIT Press Environmental Studies Series

Gordon J. F. MacDonald, consulting editor

Man, Materials, and Environment, National Academy of Sciences and National Academy of Engineering, 1973.

Citizen Groups and the Nuclear Power Controversy: Uses of Scientific and Technological Information, Steven Ebbin and Raphael Kasper, 1974.

CITIZEN GROUPS AND THE NUCLEAR POWER CONTROVERSY:
USES OF SCIENTIFIC AND TECHNOLOGICAL INFORMATION

Steven Ebbin

Raphael Kasper

The MIT Press
Cambridge, Massachusetts, and London, England

Any opinions, findings, conclusions, or recommendations expressed herein are those of the authors and do not necessarily reflect the views of the National Science Foundation (Grant: NSF GI-33432).

Library of Congress Cataloging in Publication Data

Ebbin, Steven.
Citizen groups and the nuclear power controversy.

(The MIT Press environmental studies series)
1. Atomic power-plants -- Environmental aspects -- United States.
2. Environmental protection -- United States -- Citizen participation.
I. Kasper, Raphael G., joint author. II. Title.
TD195.E4E2 621.48'1 73-20218
ISBN 0-262-05014-5 (hardcover)
ISBN 0-262-55003-2 (paperback)

"Knowledge will forever govern ignorance. And a people who mean to be their own governors must arm themselves with the power knowledge gives. A popular government, without popular information or the means of acquiring it, is but the prologue to a farce or tragedy."*

*Letter to W. T. Barry from James Madison, August 4, 1822

PUBLISHER'S NOTE

The aim of this format is to close the gap between the preparation of certain works and their publication in book form. A large number of significant though specialized manuscripts make the transition to formal publication either after a considerable delay or not at all. The time and expense of detailed text editing and composition in print may act to prevent publication or so to delay it that currency of content is affected.

The text of this book has been photographed directly from the authors' typescript. It is edited to a satisfactory level of completeness and comprehensibility though not necessarily to the standard of consistency of minor editorial detail present in typeset books issued under our imprint.

The MIT Press

CONTENTS

STUDY ADVISORY COMMITTEE

GORDON J. F. MacDONALD, Director, Environmental Studies Program, Dartmouth College (formerly member of President's Council on Environmental Quality), Chairman.

PETER BORRELLI, East Coast Representative, Sierra Club of America.

DAVID DOMINICK, Assistant Administrator for Categorical Programs, U.S. Environmental Protection Agency.

MEYER EISENBERG, Partner, Lawler, Sterling and Kent, Attorneys.

DENIS HAYES, formerly Executive Director, Environmental Action.

NORMAN KAPLAN, Chairman, Department of Sociology, Northeastern University.

JEROME KRETCHMER, formerly Administrator of New York City Environmental Protection Administration.

MARTIN LANG, Commissioner of Water Resources, New York City.

CYNTHIA WILSON, Legislative Aide, National Audubon Society.

GORDON WOLMAN, Chairman, Department of Environmental Engineering, Johns Hopkins University.

ACKNOWLEDGMENTS

All studies of this sort depend upon the generosity of others. The National Science Foundation (RANN-ERPA) granted the funds that made our study possible. Citizen group leadership and lawyers spent long hours talking to us, patiently enduring our probing and often annoying questions about their motivations, their finances, and their strategies and tactics. Scientists, working with citizen opposition intervenor groups and others with equipment manufacturers, utility companies, and the AEC, patiently described their efforts and frequently details of their means of support.

Two research assistants, John J. Zimmerman and Bruce Lederer, performed the arduous task of reading AEC documents and seemingly endless pages of hearing transcripts. Ann Finkelstein and Kristin Larson proofread this manuscript. To all of them we express our appreciation.

For some, though, to quote from Kung Fu, "to thank with words is to cheapen the gift." There are several who fall into this category, people who gave continually throughout the project their comments, their insights, their criticisms, their encouragement and moral support. Chief among them are the members of our Advisory Committee, all of whom reviewed this report at various stages of development and who, in a major sense, gave considerable impetus to its evolution and ultimate form. Their names are listed opposite. Certainly their efforts, along with those of Warren H. Donnelly of the Congressional Research Service and James A. Fay of MIT, were valuable beyond measure.

But no acknowledgment would be complete without a "without whom it would not have been possible" paragraph. Ours is short and includes only Margo von Kaenel, our Administrative Assistant who endured late hours, our dyspepsia, our frustrations, and our anxieties and stood amidst it all steady and unshakeable; and Ronald Tipton, our research assistant and lawyer-in-residence, whose pithy comments, hard work, and nose for ferreting out documents and information were invaluable assets for our small interdisciplinary team.

We do, of course, accept personally all responsibility for the entire document. We hope that what we have written makes some real contribution to what we feel deeply is essential--a democratization of governance.

Steven Ebbin
Raphael Kasper

CHAPTER I: INTRODUCTION

This report is directed towards analyzing citizen input into the nuclear power plant siting controversy. The subject is one which is characterized by complex, difficult issues which cannot be resolved by simple solutions or easy panaceas. We have looked hard and carefully at a process, essentially a public hearing process, that takes place within circumscribed rules of participation for all parties, which, we conclude, is neither fair nor expeditious, nor particularly revealing of scientific truth or technological fact. To the extent that scientific and technological issues are resolved in the process, it is despite, and not because of, an adversarial process which is essentially hostile to the goal of arriving at scientific truth. In such a process, participants seek to assure that technological facts introduced by their opponents will either support their own contentions or, at the very least, not damage their case.

The complex scenarios we have viewed are essentially exercises in which government and industry have tended to become allied against small groups of concerned, even worried, citizens. Clearly, the weight of influence, talent, money, power, policy and decisionmaking lies with government and industry. As a result, citizen groups are usually restricted to raising questions about matters concerning which they possess little knowledge or expertise.

For all intents and purposes, we found ourselves the only observers in attendance at administrative hearings with no direct stake in the outcome. This is a matter which should be of concern since a process which lacks impartial evaluation tends to become responsive to the needs and desires of only those parties with direct, often economic, interests in the outcome. It is clear, however, that the consequences of the development of nuclear power affect a constituency far broader than the nuclear industry, the Atomic Energy Commission, and even the citizen groups which intervene in licensing cases.[1] A fair and expeditious licensing process is of importance to all of us, and to our progeny.

Our study, carried out over the course of about one year, adopted what we considered to be a pragmatic methodology. We began by discussing our approach with several people who had been involved in the nuclear power plant licensing controversy, in one way or another. We spoke with officials of the Atomic Energy Commission, the Council on Environmental Quality, attorneys for citizen group intervenors and others. We reviewed newspaper files, AEC press releases, journal articles and the growing number of books on nuclear power. We then chose cases to examine in detail. The choice of cases was based in part upon the aforementioned interviews and review of the literature. We decided to study cases which represented particularly interesting examples at each stage of AEC procedures where public input to the decisionmaking process is provided for by the Rules of Practice of the Agency: the construction permit hearings on the plants proposed at Midland, Michigan by Consumers Power Company, the operating license proceedings for the Vermont Yankee Nuclear Power Plant at Vernon, Vermont, and the rule-making hearings on criteria for emergency core cooling systems. We directed ourselves, for the most part, to questions involving the Atomic Energy Commission licensing and rulemaking proceedings;

---

[1] As of March 31, there were 30 operable nuclear power plants in the United States. Another 60 plants were being built and 75 more were in the planning stage. (USAEC figures)

the activities of state, local, and other federal agencies were not given detailed attention.

Familiarizing ourselves with the docket of each case, the issues contended therein, the procedures and the actors, we proceeded to attend hearings on each of the cases. We carried out focused interviews with the participants in each case, spending as much time as feasible with each, to the extent possible in the most informal surrounding available. At each interview, at least two of us were present at all times. We attempted to keep only minimal notes during the actual interview. Later, we isolated ourselves from each other and wrote up our impressions, comparing notes and developing additional points thereafter.

During the entire course of the study we maintained repeated contacts with intervenor attorneys and citizen group members, AEC staff and Commissioners, lawyers for applicants and others with an interest in the decisionmaking process as it relates to nuclear power. We met, as well, with our Advisory Committee at several stages of the study and received and considered their recommendations which have become, both implicitly and explicitly, part of this document.[2]

In short, then, the transcripts of hearings, the docket in each case, interviews with the participants, a review of relevant literature and discussions with a broad spectrum of parties constitute the inputs to the study.

After more than one year of study, we have come to believe that our findings apply to a much broader realm of government-citizen relationship than simply to the processes of a particular agency, the AEC. We have found

---

[2] In carrying out our study, we have attended 48 days of hearings and interviewed more than 100 people. A complete list of those interviewed appears as Appendix A.

that, despite lip service paid to citizen participation in governmental decisionmaking, agency arrogance, expert elitism, stacked-deck proceedings, and the consigning of citizens to helplessness before the steamroller of big government are more the rule than the exception. On the other hand we have found strong evidence among citizen groups of "know-nothingism", blind anti-technology and anti-government sentiments, pessimism, and doom forecasting. We have not, though, viewed the problem as one of good guys vs. bad ones; we have not generally identified heroes or villains, we have not looked upon the excesses of one party as better or more justified than those of others, and we have attempted scrupulously to view the panoply of issues and the cast of characters analytically and fairly.

We have, as well, necessarily had to take account of the practical politics and economic realities which motivate nuclear proponents and have been impressed with their influence on the total process. At the same time, we have tried to understand the basis on which the citizen groups oppose a nuclear plant and have marshalled their opposition despite the evidently overwhelming odds which confront them.

We take no position concerning nuclear power as an energy source. We neither support the construction of nuclear power plants nor do we oppose their construction as a matter of general principle.

It might be noted that utility companies frequently cite the combatting of a potential energy shortage as a justification for the building and operating of nuclear power plants. Citizen groups opposing such plants often assert that the rapid increase in electricity consumption in recent years (and projected for the future) can be attributed to the policies and attitudes of the utilities and argue that the solution is to conserve electricity rather than increasing generating capacity. Thus, the debate over the future of nuclear power involves the views of various parties about the "energy crisis." Since we do not propose to enter that debate, our views of that "crisis" are not relevant to this discussion and we have attempted to keep them out of these pages.

Citizen groups have been accused by some of causing long and costly delays in the construction and licensing of nuclear power plants. There is no question that the duration of the licensing process has lengthened in recent years. But while citizen interventions have evidently contributed to delays, it is probable that misestimates of labor productivity, materials requirements, and delivery schedules, the complexity of the licensing review process, difficulties in interpreting and meeting requirements under the National Environmental Policy Act and, in some cases at least, poor planning have been the overriding factors in the slowing of the process.[3] Of course, delay tactics are at times resorted to by some citizen intervenors who see delay as the only course left to them in a system where they have little other influence on the decision-making process.

We have found that the licensing process is one which is geared to the promotion of nuclear power plants. The hearing process was originally designed to provide a mechanism to inform the local citizenry of the benefits of a nuclear power generating plant. The advent of contested proceedings has altered the role of the hearings, but it is evident to us that the decision to build remains

---

[3] For example, see: Atomic Safety and Licensing Appeal Board Memorandum and Order (ALAB-124), May 23, 1973, p. 26.

one that has already been made by the time citizens are afforded an opportunity to express their views on the matter. As this report will show in considerably greater detail, once an application for a permit to construct a power plant comes up for hearing, considerable capital expenditures have already been made and others arranged for, a site has been acquired, the staff of the AEC, the applicant and the vendor have worked out the details of construction, and the Preliminary Safety Analysis Report (PSAR) has been submitted. Not until that point, after a significantly large allocation of both financial and manpower resources has already been made, do citizens become privy to the plans. Only then does an opportunity present itself for their participation. It is at a time when hearings have been announced and a petition to intervene approved, that documents to which both the applicant for a permit or license and the AEC staff have had access for some months are made available to citizen intervenor groups. Moreover, applicants in the past have been selective in the documentation they have been willing to transmit. Citizen intervenors, then, have little advance knowledge of the full range of documents available and are forced to rely on the candor and generosity of the applicant and the AEC for whatever information is given them.

But it would be an error for us to perpetuate what seems to be a widely held misconception about the role of the AEC. The Atomic Energy Commission does not itself propose the construction of any particular nuclear generating station (with the exception of those nuclear facilities which are part of its internal research and development effort), nor does it recommend particular sites for power plants. Rather it sets the standards, criteria and rules for construction, licensing, and operation, and considers applications for permits and licenses as they relate to those standards. The individual Atomic Safety and Licensing Boards (ASLB's) which preside at licensing hearings and about which there is extensive discussion throughout this report, are not convened to determine if nuclear power plants should or should not be built. They decide only whether or not a proposed plant meets the safety and environmental criteria which the Agency has devised.

But even that consideration by the ASLB, acting on behalf of the AEC, is usually limited to the consideration of issues placed in contention (in a contested proceeding) by the intervenor(s). It follows that the burden of raising issues and proving shortcomings in safety, environmental and quality assurance programs lies with the intervenor, the Agency having been satisfied earlier in the application process that such matters had already been sufficiently provided for. According to the AEC's Rules of Practice,

> As to matters pertaining to radiological health and safety which are not in controversy, boards are neither required nor expected to duplicate the review already performed by the regulatory staff and ACRS, and they are authorized to rely upon the testimony of the regulatory staff, the applicant, and the conclusions of the ACRS, which are not controverted by any party.[4]

Normally, the participation of the AEC Staff in the hearings, at least insofar as the public is concerned, is based upon the assumption that the plans previously approved for construction (and/or operation) of the facility are sufficiently protective of the health, welfare and safety of the public and that no additional matters of safety need be raised. Opposing interventions, then, must challenge both the applicant and the AEC Staff (representing the Federal government and appearing on behalf of the public) before the licensing board.

---

[4] AEC Rules of Practice, Appendix A, Part (f), subpart (1).

The foregoing is not to deny, of course, that in a generic sense the Commission has, since the early 1960's, promoted the adoption of nuclear generating technology by utilities, and the above discussion refers only to individual nuclear power plants built by private utility companies.

Intervenors can raise issues of environmental impact, of safety, etc., but the construction and operation of the plant at that point is virtually guaranteed. Often, delay in the administrative process and challenges to that process in the courts provide citizen groups with at least the hope that they can raise the cost of the plant beyond the point at which it is economically feasible for the utility to proceed, a tactic resorted to by at least some groups.

In viewing the process at all stages of the Atomic Energy Commission licensing and rule making procedures, we have been impressed with the detailed attention given by the Agency to procedural due process and the clear lack of substantive due process afforded opponents of a permit or license. We feel that it does not help to encourage participatory democracy within a governmental structure which incorporates procedures that facilitate citizen participation but which is otherwise designed to negate the effectiveness of that participation. Strict attention to appearances of fairness and administrative propriety is, in our view, not sufficient to guarantee meaningful public involvement.

Despite these problems, citizen intervenor groups have been able to raise significant issues concerning environmental impacts and safety standards and criteria. They, through their attorneys, have probed deeply and have brought important scientific and technological issues to the surface. They have, by their interventions, induced the Atomic Energy Commission to look at issues it might not otherwise have examined publicly; they have, by their interventions, raised significant matters of nuclear science and technology which are now being exposed to public scrutiny. Problems of criteria governing emergency core cooling systems, "as low as practicable" radioactive releases, and the effects of the nuclear fuel cycle (mining, milling, manufacturing, processing, transportation, disposal) have all been exposed by intervenor groups in public forums as have numerous matters relating to compliance with the National Environmental Policy Act of 1969, particularly the impact of thermal effluents on the ecology of water resources.

Whether in fact these issues would have been raised in the absence of such voluntary, self-financed citizen interventions is, of course, academic. It is important to recognize that they had not been discussed in public before.

It should be noted that, while it would be incorrect to characterize the active intervenor groups as representative of the general population or even the popular will, it would be equally incorrect to dismiss their contentions as purely self-serving or for personal profit. Most members of the groups we have studied are not anti-technology freaks or weirdos. They are, for the most part, intelligent and upstanding members of their communities who are concerned or, perhaps, afraid of nuclear power.

EVOLUTION OF CONCERN

The first contested nuclear power licensing case began in 1956. It was brought by a labor union and directed against the construction of the Fermi fast breeder reactor near Detroit, Michigan. The construction permit for the plant was ultimately granted over intervenor objections. Appeals were taken to the Supreme Court where, in 1961, it was held in Power Reactor Development Co.

v. International Union of Electric, Radio and Machine Workers that the Commission had proceeded properly in issuing the provisional construction permit while deferring until the operating licensing stage a definitive finding that operation of the facility would provide adequate protection for the health and safety of the public. Dissenting Justices William Douglas and Hugo Black argued that the legislative history of the Atomic Energy Act made it apparent that safety issues must be resolved before a construction permit is issued. In their words,

> They [the Commission's findings] presuppose that "safety" findings can be made after construction is finished. But when that point is reached, when millions have been invested, the momentum is on the side of the applicant, not on the side of the public.[5]

Though its operation was subsequently permitted and the project was considered a valuable demonstration program, Fermi has recently been decommissioned following a series of mishaps and its failure to operate economically or to fulfill the expectations of the Commission and its sponsors.

This initial intervention was seen, at the time, as an isolated incident. A retrospective look, though makes it appear to be the beginning of what was to become a long line of contested plant sitings, a growing concern for nuclear safety, a gathering momentum of concerns for the impact of nuclear plants on the environment, and a mounting of dissatisfaction within both the AEC and its National Laboratories. At least in part because of the belief among intervenors that an alliance has existed among the Commission, the nuclear industry and the Joint Committee on Atomic Energy of the Congress, this public concern has steadily

---

[5] Power Reactor Development Co. v. International Union of Electric, Radio and Machine Workers, 367 U.S. 417 (1961).

increased. Fed by the internal debate at the Agency and at the National Laboratories, it has led to an increase in the number and intensity of active public interventions in siting and operating proceedings as well as in generic, rulemaking hearings.[6]

The early citizen interventions in the nuclear licensing process were evidently not anticipated by the AEC and were viewed by the utility industry as local matters with little implication for the national scene.

In 1961, the Pacific Gas and Electric Company (PG&E) planned to construct a nuclear power plant at Bodega Head (50 miles north of San Francisco). A small group of protesters who based their opposition on the preservation of a unique coastal site banded together to oppose this siting. The opposition grew and the initial concern for conservation was supplemented by concern for the construction of a nuclear facility on the San Andreas fault line, a center of considerable earthquake activity.[7] The Bodega Head Association released some 1,500 balloons from the site of the proposed reactor with a message attached to each which stated "this balloon could represent a radioactive molecule of strontium 90 or iodine 131." Plans to construct the Bodega Head plant were cancelled in 1964.

In December, 1962, Consolidated Edison planned to construct a plant at Ravenswood, Queens which was to be a 1,000 MWe facility located 1/2 mile from Manhattan. This turned out to be a public relations error of the first magnitude if not a potentially serious error of nuclear safety, given the huge urban

----

[6] For a more complete discussion of internal debate within the AEC, see, for example, Robert Gillette, Science, Vol. 177, September 1, 8, 1972.

[7] The AEC and concerned citizens have been at odds over the advisability of siting nuclear plants along the San Andreas fault ever since.

population of New York City. Consolidated Edison withdrew its proposal some 13 months later after a heated public dispute.

The point of real controversy, though, seemed to begin in earnest during 1969 when several anti-nuclear articles and books found their way into the reading lists of the nascent critics of the AEC and nuclear power. (A representative list of articles and books is appended to this report.) A sudden rash of literature and press articles was directed toward nuclear issues and raised serious questions about the effects of nuclear power plants. They illustrated the growing concern among some scientists and some members of the public about the implications of our nation's commitment to nuclear power and about the growing number of nuclear plants that were being built around the country. At this same point in time, the National Environmental Policy Act of 1969 was being debated in the Congress and the Administration had officially committed itself to environmental restoration. This led to a confluence of issues, the parameters of which included quality control in construction, environmental impacts, radiation exposure and its cumulative impact, and the possibility of accidents, all juxtaposed with an apparently growing public demand for electrical energy.

In September and October 1969, heated public meetings took place at Burlington and Bennington, Vermont concerning AEC plans to locate an experimental breeder reactor at Easton, New York at the Southern tip of Lake Champlain, a proposal to build a 1,000 MWe plant at Charlotte, New York (also on Lake Champlain) and the ongoing construction of a nuclear plant on the Connecticut River at Vernon, Vermont. In response to welling public opposition, the AEC sent three officials to Bennington to debate nuclear safety issues with a trio of nuclear power critics.

Before coming to Vermont, the AEC had been concerned about the crescendo of criticism leveled at it, particularly by questions raised about the safety of nuclear reactors and the impact of thermal effluents on bodies of water where nuclear plants are located. AEC officials saw a need to explain nuclear power and to deal with the questions raised by their critics in an attempt to allay public concerns. For this reason, at a meeting in Burlington a month later, one of the sessions was partly devoted to questions from townspeople gathered in the college gymnasium. Out of 130 questions received from the audience, 42 concerned effects of radiation and emissions standards, and 24 related to thermal effects.[8] The meeting was highlighted by then AEC chairman Glenn Seaborg's assertion that "with respect to resolution of the thermal effects of cooling water discharges, we do not at present have legal authority in this area."[9] His contention was based on the premise that the 1954 Atomic Energy Act gave the Commission the responsibility and the authority for the licensing and regulation of nuclear reactors from the standpoint of radiological health and safety, but did not include considering the effects of heat discharges on the receiving body of water. This was substantiated by the Supreme Court in New Hampshire v. AEC [395 U.S. 962 (1969)].

Meanwhile concern and even open attacks on the Commission and nuclear technology were growing in number and intensity, in the press, at public meetings, among environmental groups, and even in the halls of Congress. Indian Point, (New York), Palisades (Michigan), Shoreham (Long Island, New York) and Monticello

---

[8] Bill Perkins, "The U.S. Controversy - Issues of Concern", Atomic Industrial Forum, (1972) p. 4.

[9] AEC Report on "Nuclear Power and the Environment", Proceedings, University of Vermont, Burlington, Vermont, September 11, 1969, p. 94.

(Minnesota) became well known names after each of these proposed sites for nuclear power plants became the object of citizen opposition and active intervention into the licensing process provided for under the AEC's Rules of Practice.

CITIZENS AND TECHNOLOGICAL ISSUES

In a nation which had become accustomed to citizen apathy, growing protest against government actions in the 1960's and early 1970's was an increasingly evident and perhaps painful phenomenon. Protests against racial policies, against the Vietnam war, against political parties, against poverty, hunger, exclusion, neglect, and economic policies have grown in frequency and intensity. Environmental protests have likewise become more frequent and more evident and have influenced decisions on building construction, manufacturing processes, land use, off-shore oil drilling, highways, bridges, airports, power plants, housing, conservation and a broad range of public policies.

In all of these examples it would be difficult to argue that some sort of science and technology has not played a role in the process of decisionmaking. But such inputs have usually been used most effectively by the proposer of a project--a utility company, a port authority, a state, local or federal governmental agency, or an industrial organization. The proposer usually has a great financial stake in carrying a project to fruition and therefore carefully and forcefully articulates the benefits of the project applying its (usually) significant resources to the task. By contrast, groups of disparate private citizens must seek to aggregate and accrue from voluntary donations the financial resources necessary to carry their case through the uncharted administrative process and perhaps ultimately to litigation. Interventions are generally expensive, burdensome, and protracted. Although a decision upholding the contentions of intervenors might result in small changes in design or operating characteristics of the project, the odds are heavily against a cancellation of the project.

Citizen intervenors have little money and very little manpower skilled in scientific or technology analysis or environmental impact assessment. Technology based industries and technology based government agencies, like the AEC, already have on hand numbers of scientists and engineers either directly employed, experienced as consultants, or university based and therefore reliant upon them as outside sources for research funding. Large industries (e.g., equipment vendors and utilities) and certainly most grant giving agencies (e.g., AEC) have a natural following among academics who rely on them for research funds and graduate assistantships, and who thus can be called upon for support. Virtually all worthy nuclear scientists and/or engineers have some ongoing relationship to the AEC or the nuclear industry which would be jeopardized by their identification with opponent intervenor groups. And to the extent that competent and experienced scientists willing to identify themselves with intervenor's contentions are available, they are difficult to locate and expensive to engage.

Moreover, the scientific and technological issues involved in nuclear power cases are infrequently so sharply drawn as to allow most scientists and engineers to deal with technical issues falling strictly within their areas of expertise. Honest scientific and technological testimony would necessarily admit to gaps in available knowledge and to uncertainties about certain aspects of the effects of nuclear power plants. In the politically charged atmosphere of the adversary process, a clever lawyer tends to seize upon such admissions of a lack of omniscience as attempts to hide potentially damaging evidence. In addition, most scientists and engineers possess certain discrete areas of expertise. The areas of concern in nuclear power cases, however, often involve many interacting issues which tend to go beyond the expertise of any single technical expert. Because it is so easy for a lawyer to embarrass a witness by questioning in areas near but not within his or her field of expertise, the role of the technical witness in the adversary proceedings on nuclear power cases is one which many scientists and engineers try to avoid. Still another reason for the discomfort of technical experts in adversary proceedings arises because many scientists and engineers find it difficult to express scientific or technological concepts or opinions in a manner understandable to laymen, including lawyers, something which they are called upon to do constantly during contested hearings.

Our research indicates that the choice of scientific assistance by intervenor groups usually depends on the willingness of individual scientists to advocate a particular position categorically and stoically. More frequently than not, the intervenors must rely on people from disciplines peripheral to the central issues of concern, scientists whose publications are judged by their peers to be of questionable scientific merit, or inexperienced graduate students. In some few cases, intervenors obtain the services of outstanding scientists. But independent scientific analysis and evaluation is almost never available.

Furthermore, it is clear that normal scientific processes have little, if any, relationship to the adversary proceedings we have witnessed. It is the lawyers who are the advocates on behalf of their clients in the hearing process; by training, background, and legal practice the lawyer's role is not necessarily to get at the truth but rather to use the legal process to develop and win the case for his side and often to obscure matters of fact when those facts do not tend to substantiate his arguments. Moreover, few lawyers in our investigations demonstrated any real understanding of science or engineering. Some had better understanding than others, but few if any attorneys with whom we spoke had any substantial background in science or engineering. Those who claimed any such training usually alluded only to a couple of courses taken as undergraduates. Some, however, have taught themselves enough of the technical phraseology and engineering jargon so as to be able to effectively probe for soft spots in witnesses' testimony.

MOTIVATION FOR CITIZEN INVOLVEMENT IN NUCLEAR POWER ISSUES

Citizen intervention in nuclear power plant siting cases can be explained, in part, as a natural outcropping of and in some respects a culmination of a more generalized national concern with environmental degradation. Nuclear power plants require sites contiguous to bodies of water since vast supplies of water are essential for operation of the plants. They are then in competition for a limited environmental resource with other human uses or values, often recreational or esthetic.

But the concerns that induce opposition to a nuclear power plant are based upon more than simply environmental values. Fear of things nuclear or atomic is not uncommon among the population in general and is rather evident among intervenor groups in particular: fear of nuclear accidents, fear of radiation exposure, fear of propinquity to nuclear plants, fear of long-range unknowns, fear of technological error and of commitment to a technology imperfectly understood.

This fear has an evident historical foundation. Most adult Americans were first made aware of nuclear power as a force for destruction. Our introduction to nuclear power came with the bombing of Hiroshima and Nagasaki. Were this beginning not sufficient to cast a lasting pall over the technology, the American public was almost incessantly reminded of the danger of nuclear attack throughout the Cold War years of the 50's, which were characterized by "brinksmanship" diplomacy. Nuclear confrontation over the Cuban Missile crisis in 1962 brought this message home even more clearly. Only in recent years have the major powers come to treaty terms on the testing and control of nuclear weapons; yet the use of nuclear explosives for testing purposes continues even today, almost always accompanied by strident opposition and expressions of concern. The image of the mushroom shaped cloud, which bears no realistic relationship to nuclear power plants, remains clear and frightening in the minds of the American public.

Furthermore, the development of nuclear power has been marked by several nuclear "incidents" and these incidents continue to occur. Reactors, after all, are man-designed and man-fabricated and accidents are possible if not highly probable. But the safety record of the nuclear power generating industry, while not without blemish, has been by all odds outstanding, and to date no nuclear accident at a power plant has led to fatalities among the general population.

Although the majority of the population in the particular areas of existing or planned reactors we have studied has not actively participated in the interventions, it is not clear that this quiescence has represented a full or unquestioning acceptance of nuclear power. Rather, as public opinion polls and our observations reveal, many people associate "nuclear" with anxiety (and with words such as fear, danger, destruction, rac health, death, burn). A recent survey pointed out that 60% of adults associate the word "nuclear" with violence/anxiety; 83% of adults and 90% of youth associated the word "atomic" with violence/anxiety. Moreover, this study, carried out for and funded by the utilities industry, comments that "the anxious/non-anxious responses to these terms (nuclear, atomic) suggest that basic anxiety, on the part of many people, is an underlying cause for resistance to nuclear power plants."[10]

The same study goes on to state that "the main reasons offered in support of nuclear power plants are less air pollution, need for better power supply, and the hope of low cost power." It pointed out that:

> Opposition is based primarily on concern about the safety of a nuclear power plant, only secondarily on environmental considerations. Concern about safety is expressed in general terms, or regarding possible release of radioactive substances, less often in regard to possible explosion. (However, as brought out in further questioning, concern about possible explosion is about equal to concern about radioactive substances.)
>
> . . . [T]he following proportions indicate a substantial degree of negative thinking on the safety of nuclear power plants:
>
> - 52% of adults have some fear that nuclear plants might explode
> - 50% are concerned about radioactive substances being released in dangerous amounts
> - 44% are not convinced of the safety of workers within nuclear plants.[11]

----

[10] Nation Wide Public Opinion Survey, "Urban Customers and Teenagers (15-17)" served by Investor-owned Electric companies, conducted during April 1972, Electric Companies Public Information Program, Central Surveys, Inc., Shenondoah, Iowa.

[11] _Ibid_.

While we do not necessarily aver to the accuracy of the statistics presented in the survey, we do feel that they substantiate our own observations, which are less systematically arrived at. Moreover, other surveys we have seen would indicate the general validity of the conclusions reached by the study.

Looked at in this context, it is not particularly surprising that a technology identified in the public mind as having emerged from wartime weapons development has not won full acceptance among the population for its peaceful uses.

We consider the point sufficiently made that a substantial portion of the population in general and the active intervenors in particular are afraid of or substantially concerned about the dangers of catastrophic nuclear accident, (the particular kind is not terribly important for our purposes) and the danger of exposure to radioactive releases (short term as well as cumulative) and view negatively the imposition of a nuclear power plant in their neighborhood for reasons of environmental and aesthetic values. Suffice to say that an announcement by the AEC, or a utility, of a proposed nuclear facility is adequate to stimulate negative reaction among a portion of the population of the affected community, and in some areas of the country, most notably the Northeast, the Midwest, the Far West and the Northwest, this has manifested itself in numerous active interventions in the AEC construction permit or operating licensing hearing processes.

Our considerable and lengthy conversations with individuals and groups involved in interventions and our attendance at their meetings has led us to see the interventions as expressions of fear and concern sufficiently strong so as to motivate the participants to make contributions ranging from ten dollars to \$7,000-\$8,000 with a great many people contributing \$100-\$500. (Few, of course, give as much as \$5,000-\$8,000). Many also donate to the cause a great deal of their personal time and resources, attending meetings and hearings, collecting funds, clipping newspapers, as well as providing typing and other backup support services to their lawyers.

It is not difficult to make the transition from the more generalized concern with things "atomic" or "nuclear" to the more particularized concern with nuclear power for the generation of electric energy. The placement of nuclear facilities near population centers brings the more general concern into focus. No one, and this seems true of supporters as well as opponents, actively wants a nuclear power plant in his backyard, though it is also true that various inducements to communities and individuals by utilities have worked to stem some of the opposition and to make a proposed plant more palatable to a community. In the case of the Vermont Yankee Nuclear Power Plant, for example, the people of Vernon, Vermont were evidently more receptive to the siting of the nuclear reactor in their town in light of the sharp reduction in property taxes that was promised and delivered as a result of the tax contribution of Vermont Yankee. Last year, for example, Vermont Yankee paid 91.2% of the town's total property tax bill.[12] Moreover, in addition to the dramatic impact on the individual's annual tax payment, a new library and town hall and a community swimming pool were constructed.

In Midland, Michigan the inducements for community support included a rumor that if two proposed nuclear plants were not constructed, Dow Chemical Company,

---

[12] Brattleboro Reformer, March 8, 1973, p. 1.

<u>the</u> major employer in the area and the potential recipient of process steam from one of the nuclear units, would likely move its plant and corporate headquarters to a location where the business climate was more receptive to their needs. Dow later issued a denial of the "rumor" and indicated that the size of their investment in Midland would evidently not permit such a move. The object lesson, though, was made clear. There is evidence to suggest that some people who previously had taken no active position in favor of the plant became more actively supportive of it and considered their jobs, and, therefore, their livelihood to be threatened by the citizen group interventions. One apparent result of this fear of economic disaster which could result from a denial of the construction permit for the plant has been the harassment of at least one of the leaders of the intervenor groups by other members of the community. Proprietors of small businesses, workers, farmers, brine wellmen and others whose main source of income depended upon the presence of Dow Chemical, were stirred sufficiently by the project to lend their support for the plant.

Despite these inducements to support the construction and operation of nuclear power plants, it is clear that substantial portions of the public are quite concerned about the presence of such power plants in their communities. But while the number of contested nuclear power plants at both the construction and operating licensing stages have increased dramatically since 1969, our observations have led us to believe that they are, for the most part, essentially individual, local actions and do not represent a consolidated national effort, as is often alleged. The people who may oppose the construction of a nuclear facility in California, for example, are likely to have little, if any, active role or participation in an intervention taking place in Vermont. And, though there may be a certain degree of sympathetic identification with other opposition intervenor groups, it is only at the level of generic rulemaking hearings that any consolidated effort has manifested itself.[13] In our view, then, opposition interventions are for the most part local(ized) undertakings. The appearance of a handful of lawyers in many interventions does, however, serve to broaden the interrelationships among the various cases. However, this in no way suggests that the opposition is a nationally concerted one since the impetus and funding for each intervention are essentially of a local nature. There is even some evidence that suggests to us that were plans to build such a plant in one locality to change and a different site elsewhere selected, those who have joined the opposition would drop their active roles.

NUCLEAR SAFETY

Just as its destructive force was heralded in the late 1940's-1950's, so too were the peaceful applications of nuclear power subjected to public relations-type hyperbole about the "gee whiz" wonders that nuclear power would bring to peacetime concerns. Such claims were given legitimacy by the passage of the Atomic Energy Act of 1954, by the support of peaceful uses of atomic energy by the Joint Committee on Atomic Energy on behalf of the Congress, and by the promotional activities of the Executive Branch of the government through the Atomic Energy Commission. Through these governmental institutions a variety of subsidies were made available to the equipment manufacturers and to utility

---

[13] Certain national environmental groups, such as the Environmental Defense Fund and the Natural Resources Defense Council, have joined individual cases as opponent co-intervenors. EDF, though, has tended to be concerned with assuring that the procedural requirements of the National Environmental Policy Act of 1969 are adhered to, while NRDC has usually directed its interests to questions of nuclear fuel transportation and the disposal of nuclear wastes.

companies, including public assumption of research and development costs, the Price Anderson Act of 1957 (which limits liability for any nuclear power plant accident to a total of $560 million, including $465 million in public funds) and other financial inducements to utilities to "go nuclear."

A further spur to the adoption of nuclear technology for the production of electricity was the Private Ownership Act of 1964, which enabled utilities, after a transitional period, to own the nuclear materials essential for the production of reactor fuel.

In addition, growing public concern with air pollution from the stack emissions of fossil-fueled plants (and other sources), diminishing supplies of natural gas, issues of environmental degradation resulting from surface and underground mining of coal, rising prices of coal, oil, and gas, and mining safety issues as well as diminishing supplies of domestic oil and increasing reliance on foreign sources made the transition to nuclear power more and more attractive to the utilities industry and to the government. The transition was expedited by the federal government, which convinced the utilities that nuclear power would provide a more stable, economically competitive and cleaner source of electrical energy. At first there was reluctance among power companies to accept nuclear technology, given its yet unresolved problems and the large capital outlay and investment entailed in construction of nuclear facilities. However, a willingness of vendors to use the initial plants as "loss leaders" and federal government underwriting combined to induce an industry commitment to nuclear power.

Public confidence in the reliability and safety of nuclear power plants, though, was not assisted by the dual role of the AEC for promotion and regulation of nuclear power development provided for in the 1954 Atomic Energy Act, nor bolstered by the catalogue of incidents that occurred over the few years of commercial operation of nuclear reactors for production of electricity. Repeated claims of almost absolute safety were not substantiated by such incidents that spelled to many the inevitability of public exposure to a nuclear accident as more and larger reactors were built across the nation. Though the AEC and utilities claim that no member of the general public has ever lost life because of a nuclear power reactor accident, claims of absolute safety were not substantiated by the record or by the increasing stridency of internal debate within the "nuclear community" that began to leak out to the public.

In 1957, for example, an AEC report, WASH-740, based on research done at the Brookhaven National Laboratory (AEC) estimated the "(1) possible consequences of major accidents and (2) the likelihood of occurrence of such major reactor accidents."[14] WASH-740 essentially attempted to quantify the potential for large scale nuclear reactor accidents and concluded as follows:

> 1. Many power reactor systems will operate under high pressures. High pressure systems are subject to failure.
>
> 2. The cumulative effect of radiation on physical and chemical properties of materials, after long periods of time, is largely unknown.
>
> 3. Various metals used in reactors such as uranium, aluminum, zirconium, sodium and beryllium, under certain conditions not at present clearly understood, may react explosively with water, also present in many reactors. . . . Chemical reactions of enough violence to rupture the containment vessels, with release of fission products, could occur.

---

[14]"Theoretical Possibilities and Consequences of Major Accidents in Large Nuclear Power Plants," AEC Document WASH-740, 1957, p. viii.

> 4. After initial operation, many of the vital components become inaccessible for inspections. In non-nuclear plants, serious accidents are often averted through detection of incipient failure.
>
> 5. Much remains to be learned about the characteristics of nuclear systems.[15]

For three types of assumed accidents, theoretical estimates indicated that personal damage might range from a lower limit of no injuries or fatalities to an upper limit of about 3,400 killed and about 43,000 injured. Theoretical property damage ranged from one-half million dollars minimum to a maximum of around seven billion dollars. It was estimated that people could be killed at distances up to 15 miles, and injured at distances of about 45 miles.[16]

WASH-740 is, of course, now 16 years old. Present power reactors are more sophisticated and considerably larger than those which formed the basis for the 1957 accident report. There has been no publicly issued updating of WASH-740, a fact which in and of itself may be one of the causes for much of the public uncertainty about the state of nuclear power safety.

RECENT EVENTS

As the number of interventions have grown since 1969, the repetition of issues raised by the opposition became more obvious. In 1971, the AEC responded by excising the issue of radiation emission standards from individual power plant siting hearings and made that issue a subject of generic, or rulemaking hearings. Subsequently, a like decision was made by the Commission in questions raised by intervenor groups about Emergency Core Cooling System criteria and

----

[15] *Ibid.*, p. 5.

[16] *Ibid.*, p. viii.

issues relating to the nuclear fuel cycle which became the subject of rule making hearings as well.

In addition to these public hearings, pressure brought by intervenor groups, notably the Union of Concerned Scientists (Cambridge, Massachusetts), Businessmen for the Public Interest (Chicago, Illinois), and Ralph Nader, under the provisions of the Federal Advisory Commission Act of 1972 (PL 92-463), forced the AEC under threat of suit to open to the public at least part of the heretofore closed meetings of the Advisory Committee on Reactor Safeguards (ACRS) beginning March 8, 1973. Proprietary information is still to be kept secret under the new rules, but a considerable portion of the meetings, including the interchange between ACRS, AEC Staff and the applicant, is now open to the public.

The National Environmental Policy Act of 1969 became law on January 1, 1970. Clearly, it has had a profound impact on the decision making responsibilities of all Federal administrative agencies whose activities have a major effect on environmental quality, including the Atomic Energy Commission. The purpose of the Act was

> to declare a national policy which will encourage productive and enjoyable harmony between man and his environment; to promote efforts which will prevent or eliminate damage to the environment and biosphere and stimulate the health and welfare of man; to enrich the understanding of the ecological systems and natural resources important to the Nation; and to establish a Council on Environmental Quality.[17]

Section 102(2)C of the Act requires that all agencies of the Federal Government prepare, in connection with proposals for legislation and other major Federal action which significantly affects the quality of the human environment,

---

17 43 USC 4321, et seq.

detailed statements on (i) the environmental impact of the proposed action; (ii) any adverse environmental effects which cannot be avoided should the proposal be implemented; (iii) alternatives to the proposed action; (iv) the relationship between local short-term uses of man's environment and the maintenance and enhancement of long-term productivity, and (v) any irreversible and irretrievable commitments of resources which would be involved in the proposed action should it be implemented.

Three distinct legal issues involving the agency's preparation of this impact statement have been the subject of dozens of court cases under NEPA.[18] In the first place the courts have often been asked to determine whether a specific project involves a "major federal action significantly affecting the quality of the human environment," and thus requires that an environmental impact statement be filed. Courts have upheld agency decisions that "102 statements" were not required for a military practice maneuver in Reid State Park in Maine, for erecting a Federal office building in Mobile, and for grants to assist construction of a 66-unit apartment project in Los Angeles. On the other hand, they have required statements for Interstate Commerce Commission approval of a temporary boost in railroad freight rates,[19] for Federal aid to widen a Wisconsin State highway, and for a Soil Conservation Service project to rechannel 66 miles of Chicod Creek in North Carolina.[20]

---

[18] Over 150 NEPA-related suits had been litigated in federal courts as of the end of 1972.

[19] This lower court decision was overturned by the Supreme Court in a June, 1973 decision, one of the first cases reviewed by the High Court involving interpretation of NEPA. See Scrap v. ICC, decided June 17, 1973. In the same case, however, the Court agreed that the plaintiff group of law students at George Washington University had standing to challenge final ICC rate regulations under NEPA based merely on their general interest in halting further environmental degradation.

[20] Council on Environmental Quality, Third Annual Report (1972), p. 252.

A second issue that has often arisen involves the content of an environmental impact statement. Courts have generally applied a standard of reasonableness in determining whether a 102 statement addresses each of the requirements of NEPA. This consideration is usually limited to deciding whether or not the decisionmakers and the public have been sufficiently informed. As the U.S. Court of Appeals, D.C. Circuit recently observed, "So long as the officials and agencies have taken the 'hard look' at environmental consequences mandated by Congress, the court does not seek to impose unreasonable extremes. . . ."[21]

The third area of inquiry of the courts concerns aspects of the actual environmental impact of the proposed action of the agency which the impact statement is required to assess. By and large the courts have limited their review in this respect to whether the agency has given adequate consideration to environmental values before taking action to initiate a given project. The substance of a decision is normally reviewed by the court only to determine if it is evidently arbitrary or capricious or to determine whether a clear mistake in judgment has been made.

Thus, most court decisions in this area to date have primarily considered the procedural requirements of NEPA. It should be pointed out, however, that a series of recent decisions has recognized NEPA as an expression of judicially enforcible _substantive_ rights, under which decisions by Federal agencies may be reviewed in court to decide if they substantially comply with the Congressionally declared goals in the statute. A recent opinion by a particularly outspoken Federal judge declared that

----

[21] _Natural Resources Defense Council_ v. _Morton_, 3 ERC 1558 (1972).

> This court is fully aware that many cases have held that NEPA does not create 'substantive' rights, but rather creates only 'procedural' obligations. To the extent that NEPA does not articulate acceptible levels of air, water, heat and noise pollution, this is an accurate position. . . . But to the extent that it would allow the agencies merely to disclose the likely harm without reflecting a substantial effort to prevent or minimize environmental harm, it is not an accurate position of the role of the courts under NEPA.[22]

Perhaps the most dramatic change in the Atomic Energy Commission's attitude toward intervenor groups took place as a result of the Calvert Cliffs court decision in 1971. There the Court of Appeals for the District of Columbia found that the Commission's interpretation of NEPA improperly limited consideration of environmental values in the Commission's decision-making process. In Justice Skelly Wright's words, "We believe that the Commission's crabbed interpretation of NEPA makes a mockery of the Act."[23] This decision required that (1) NEPA be implemented for all licensing proceedings begun after January 1, 1970, and that this implementation include independent AEC assessment of water quality and other environmental factors at every important decision making stage, including a case-by-case cost-benefit assessment balancing environmental and non-environmental factors; (2) the licensing boards of the AEC (ASLB's) must give independent review to all NEPA matters in uncontested as well as contested cases; (3) the AEC must give prompt NEPA consideration to facilities for which permits and licenses were issued after January 1, 1970, if such matters were not substantially considered in the original licensing action; and (4) with respect to construction permits issued before January 1, 1970, the AEC must consider any

----

[22] Sierra Club v. Froehlke, 5 ERC 1065 (February 16, 1973).

[23] Calvert Cliffs Coordinating Committee v. AEC, 2 ERC 1779 (D.C. Circuit, decided July 23, 1971).

significant non-radiological environmental impact and order such facility operations as may be needed. The Commission did not appeal the decision, and responded by altering its environmental review procedures to meet the requirements of NEPA as interpreted by the Court of Appeals.

(In the section on the AEC licensing process, the Commission's environmental impact assessment procedures are discussed further.)

Further changes in the Commission's rules have been made subsequently. The AEC claims that its new Rules of Practice, adopted in July, 1972 accommodate more meaningful citizen participation. But all of the rule changes, in the view of the citizen groups, are strictly addressed to procedural matters and continue to assure the AEC's desired goal: the licensing and construction of nuclear power plants as expeditiously as possible. Nothing in the new rules or the old, they contend, affords them the opportunity to have a real voice in that decision, which is essentially made by the utility and the Commission Staff even before hearings are announced. Despite these criticisms, it is somewhat ironic to note that the AEC has instituted (since 1969) clearer and more orderly procedures (or at least a process with a more evident focus) to accommodate interventions by citizen groups than have other Federal agencies with major environmental responsibilities.[24]

----

[24] While the draft of this report was in preparation, significant changes in the AEC structure were announced and still further changes were rumored. One major criticism of the Atomic Energy Commission had long been directed toward the placing of light water safety programs within the Division of Reactor Development and Technology of the AEC. In an effort to separate safety programs from its promotional and developmental programs. The AEC recently announced the creation of a Division of Reactor Safety Research reporting directly to the General Manager of the AEC. The AEC announced that "Separation of these functions from developmental activities and placement at this level will assure independent reactor safety research." [Nucleonics Week, Vol. 14, No. 20, May 17, 1973, p. 1.]

In June, 1973 the AEC began a new research program in the Directorate of Regulatory Standards. Nucleonics Week noted that "the formalized arrangement

Our interpretation of the motivation of citizen group intervenors leads us to conclude that they relate primarily to feelings of powerlessness before the decisionmaking processes of the Federal government. Though intervention is expensive and laborious, frustration with the inability to stem an apparent tide of technological determinism, fear of nuclear technology, and anger with the trade-off of diminishing natural resources for economic growth compels these citizens to make their voices heard. The physical ecology is a major issue; but the human ecology, the intangibles of quality of life, the realization that having more may mean enjoying less, and headlines about technology-related problems such as TFX and C-5A cost overruns, as well as daily articles in the press about corruption in high places, also contribute to their concern. Perhaps, only perhaps, the basic issue amounts to considerably more than merely interventions in local nuclear power reactor siting cases--perhaps it is really a crisis of governance.

----

replaces an ad hoc one in which regulatory passed along its R&D needs to the former Director of Reactor Development and Technology." [<u>Nucleonics Week</u>, Vol. 14, No. 24 (June 14, 1973) p. 1.]

Recent rumors hold that a major reorganization of the Atomic Energy Commission and, indeed, of the entire government structure for the setting and implementation of energy policy is imminent. The restructuring of the AEC would include, according to the rumors, a complete separation of regulatory and developmental responsibilities. ["AEC is Set for Split Into 2 Agencies", <u>The Washington Post</u> (June 15, 1973) p. C9.]

CHAPTER II: AEC PROCEDURES

LICENSING

In comparison to other older federal agencies, the regulatory functions of the Atomic Energy Commission have been relatively small, with the number of regulations promulgated and cases decided relatively few. Recently, however, this situation has changed drastically, as the conversion to nuclear power as a source of electricity has produced a concomitant increase in the number of reactor facilities subject to AEC licensing procedures. As of the end of 1972, 29 commercial nuclear power plants were operable, 55 were under construction, and another 76 were planned.[1]

The regulatory process in reality commences prior to the beginning of formal licensing hearings by the Commission. The utility seeking to construct a plant generally makes preliminary contact with the AEC and various state regulatory agencies before it even announces its intention to build a nuclear power station; this informal discussion with the Commission is explicitly authorized by its own Rules of Practice.[2]

Sometime thereafter the utility files a formal application for all licenses needed to eventually obtain approval to operate its reactor facility. For example, a typical license application begins in the following manner:

> Pursuant to the Atomic Energy Act of 1954 and the Commission's Rules and Regulations, Vermont Yankee Nuclear Power Corp. hereby applies for all licenses necessary for the construction and operation by it of a nuclear power plant. . . . Licenses of the

----

[1] U.S. Atomic Energy Commission, 1972 Annual Report to Congress, Regulatory Activities, p. 6.

[2] 10 CFR §2.101a.

> following types are requested: Class 104 facilities license and construction permits, special nuclear material license and by-product material license.[3]

The Atomic Energy Act requires that

> Each application for a license hereunder shall be in writing and shall specifically state such information as the Commission, by rule or regulation, may determine to be necessary to decide such of the technical and financial qualifications of the applicant, the character of the applicant, the citizenship of the applicant, or any other qualification of the applicant as the Commission may deem appropriate for the license. . . . The Commission may at any time after the filing of the original application, and before the expiration of the license, require further written statements in order to enable the Commission to determine whether the application should be granted or denied or whether a license should be modified or revoked.[4]

The key requirement is that

> the applicant shall state such technical specifications, including information of the amount, kind, and source of special nuclear material required, the place of the use, the specific characteristics of the facility, and such other information as the Commission may . . . deem necessary in order to enable it to find that the utilization or production of special nuclear material will be in accord with common defense and security and will provide adequate protection to the health and safety of the public.[5]

Initially it should be pointed out that obtaining permits and licenses is an expensive process. According to the 1972 AEC Annual Report,

> Under the new schedule, which became effective May 25, facility licensing fees were increased substantially. For example, for a power reactor plant producing approximately 1,000 MWe, the fee required for submission of a construction permit application was increased from $25,000 to $70,000; the construction permit fee from $135,000 to $300,000; and the operating license fee from $255,000 to $400,000. The annual fee for operating the plant was raised from $6,000 to $36,000. The maximum fees for

---

[3] License Application, In the Matter of Vermont-Yankee Power Company, AEC Docket No. 50-271 (November 30, 1966) p. 1.

[4] 42 USC §2232 (1954).

[5] *Ibid.*

> licenses for possession and use of special nuclear material in plutonium fuel fabrication plants increased from $8,000 to $27,000.
>
> License fees paid to the AEC during 1972 totaled $6,330,004 bringing to $9,200,893 the amount collected since fees were imposed in 1968.[6]

The application by the utility is first considered in order to determine if the applicant is to be granted a construction permit. The utility must file a preliminary safety report (PSAR) for the agency to review. This voluminous document includes a description of the site and of the proposed reactor, with particular emphasis on the features of design and construction which are pertinent to an appraisal of the safety of proposed operations; a preliminary analysis and evaluation of the safety of proposed operations; a preliminary plan for the applicant's organization, training of personnel, and conduct of operations; a discussion of procedures and devices which are to come into play in the event of an accident or other emergency; a description of the quality assurance program to be implemented at the facility; and an identification of those structures, systems or components which require research and development to confirm the adequacy of their design.[7]

Application for construction permits are also required to include

> a description of the preliminary design of equipment to be installed to maintain control over radioactive materials in gaseous and liquid effluents produced during normal reactor operations. . . . The application shall also identify the design objectives, and the means to be employed, for keeping levels of radioactive materials in effluents to unrestricted areas as low as practicable.[8]

----

[6] U.S. Atomic Energy Commission, 1972 Annual Report to Congress, Regulatory Activities, p. 25.

[7] 10 CFR §50.34.

[8] 10 CFR §50.34a(a).

Each construction permit application also must include an estimate of the amount of each of the principal radionuclides expected to be released annually in gaseous effluents to unrestricted areas during normal reactor operations, and a description of provisions for packaging, storage, and offsite shipment of radioactive materials in the form of solid waste.[9]

The staff of the Office of Regulatory Activities evaluates the safety aspects of the application with the assistance of such other elements of the AEC staff as may be appropriate. Generally there is a finding that the application lacks adequate information in one or more respects, or differs with some aspect of the applicant's evaluation. The staff attempts to resolve these differences through conferences with the applicant and by addressing one or more letters to the applicant in order to present the questions which it wishes to have answered.

The submission of this additional information is followed by informal conferences between representatives of the staff and the utility, the nuclear supply system manufacturer, and others as necessary to iron out differences as much as possible.[10] The staff must make a determination that, taking into account the specific characteristics of the facility and its location, no credible accident (including the maximum credible accident) would result in releasing radioactive material into the environment in excess of limits established in the AEC's radiation protection regulations.[11] This "safety evaluation" report becomes a part of the public record.

----

[9] 10 CFR 50.34a(b)

[10] Sidney G. Kingsley, "The Licensing of Nuclear Power Reactors in the United States", Atomic Energy Law Journal, Vol. 7 (Fall, 1965) p. 319ff.

[11] Harold Green, "Safety Determinations in Nuclear Power Licensing: A Critical View", Notre Dame Lawyer, Vol. 43 (1968) p. 641.

When the regulatory staff study is nearly completed, the application is reviewed by the independent statutory Advisory Committee on Reactor Safeguards. The ACRS consists of a maximum of fifteen members appointed by the Commission for terms of four years each, with one member designated by the Committee as its chairman. Its general function is to "advise the Commission with regard to the hazards of proposed or existing reactor facilities and the adequacy of proposed reactor safety standards."[12]

The PSAR is usually provided to the ACRS within a few days after it is submitted by the applicant; shortly thereafter, a subcommittee of four or five ACRS members is assigned to evaluate the PSAR. The role of this subcommittee is to insure that sufficient information has been developed for a committee decision, and to identify those new or novel aspects of a proposed design that warrant special consideration.

About a month after the PSAR is received, the Regulatory Staff provides the Committee with a preliminary evaluation of the project, which is not made public, identifying those matters which appear to require special attention. This report as well as the report of the ACRS subcommittee forms the basis for Committee consideration of a project.[13]

Both the subcommittee and the full committee have one or more informal meetings with the staff and the applicant. Until recently these meetings were held in private, and the minutes were not made a part of the public record. On March 8, 1973, however, the AEC opened up parts of the full Committee's

---

[12] 42 USC §20.39.

[13] AEC Licensing Procedure and Related Legislation, Hearings before the Joint Committee on Atomic Energy, 92nd Congress, 1st Session (1971) Part 1, pp. 95-96.

meetings to the public, pursuant to the recently enacted Federal Advisory Committee Act.

The ACRS report to the Commission is in the form of a letter addressed to the Chairman of the AEC. It expresses the Committee's opinion as to whether the proposed reactor can be constructed at the site with reasonable assurance that it may be operated without "undue risk" to public health and safety, and it identifies aspects of design which the Committee feels should be given special attention or require further review. The report is later introduced at the hearing with the staff's safety evaluation, and the staff takes the report into account in making its own final evaluation.[14]

Prior to the issuance of a construction permit or an operating license for a nuclear power plant, the AEC is also required to determine the potential environmental impact of that plant in order to ascertain that the action of the Commission in approving the facility will be consistent with the goals set forth by the National Environmental Policy Act. Each applicant for a construction permit must submit, separate from its application, 300 copies of a document entitled "Applicant's Environmental Report--Construction Permit Stage", which has to include:

> (a) the environmental impact of the proposed action;
> (b) the adverse environmental effects which cannot be avoided should the proposal be implemented;
> (c) alternatives to the proposed action;
> (d) the relationship between local short-term uses of man's environment and the maintenance and enhancement of long-term productivity; and
> (e) any irreversible and irretrievable commitment of resources which would be involved in the proposed action should it be implemented.[15]

---

[14] Kingsley, pp. 322-323.

[15] Guide to the Preparation of Environmental Reports for Nuclear Power Plants, Atomic Energy Commission, Directorate of Regulatory Standards, August, 1972, pp. 1-2.

According to the AEC's Draft Guide to the preparation of environmental reports, this report should include a cost-benefit analysis which balances the environmental effects of the facility and the alternatives available for reducing or avoiding adverse environmental effects, as well as the environmental, economic, technical and other benefits of the facility. To the fullest extent possible, this cost-benefit analysis must quantify the various factors considered. The report must also include a discussion of the status of compliance of the facility with applicable environmental standards and requirements which are imposed by Federal, State, and regional environmental protection agencies.

The report by the applicant, relevant published information, and any comments received from interested persons are considered by the Commission's regulatory staff in preparing a "Draft Detailed Statement of Environmental Considerations" concerning the proposed licensing action. The staff's Draft Statement along with the applicant's Environmental Report are sent for comment to the Federal Council on Environmental Quality, to certain Federal agencies, and to the Governor or appropriate state and local officials authorized to develop and enforce environmental standards for an affected state.

Comments are requested within a specified time interval on the applicant's Report and the Draft Statement.[16]

The Regulatory Staff then considers the comments received from the various Federal, State and local agencies and officials, from the applicant, and from private organizations and individuals, and prepares a "Final Detailed Statement on the Environmental Considerations." The final statement is sent to the Council on Environmental Quality and is made available for public scrutiny.

----

[16] Ibid., p. 3.

Subsequent hearing and action on environmental aspects involved in the Commission's issuance of a construction permit are based on the Commission's Final Environmental Statement. The proposed procedures require that this statement take into account information from many sources, including the applicant's Environmental Report and supplements and the comments received from the other sources mentioned above.[17]

The AEC must also make a finding that issuance of a license to an applicant will not violate anti-trust provisions of the Atomic Energy Act.

For this purpose the Commission transmits to the Attorney General a copy of the license application, and requests advice as to whether the proposed license would create or maintain a situation inconsistent with the anti-trust laws.[18] If the Attorney General considers it necessary, the Commission holds public hearings on this subject; normally, no such hearing takes place.[19]

Prior to the issuance of a construction permit, the AEC is required to hold a mandatory public hearing. This hearing does not commence until at least 30 days advance notice has been given to the public. The notice of hearing sets forth the issues to be determined in deciding whether the proposed construction permit should be granted and the time and place of the hearing. It also provides for a prehearing conference, and specifies when petitions for leave to intervene are to be filed, and how persons who wish to express their

---

[17] Ibid.

[18] 10 CFR §50.42b.

[19] Recent events indicate that anti-trust hearings are likely to become quite common as the increase in the percentage of electricity supplied by nuclear power stations results in more requests for these hearings by municipalities concerned with allocation of electricity by major power supply companies.

views on the subject matter of the hearing can do so without becoming full intervenors.

The AEC rules permit

> any person whose interest may be affected by a proceeding and who desires to participate as a party . . . (to) file a written petition under oath or affirmation for leave to intervene. . . . The petition shall be filed not later than the time specified in the notice of hearing, or as provided by the Commission or the presiding officer. . . . Nontimely filings will not be entertained absent a determination by the Commission or the presiding officer that the petitioner has made a substantial showing of good cause for failure to file on time. . . .[20]

The Commission is directed to consider the following in ruling on a petition to intervene: (1) the nature of the petitioner's right under the Act to be made a party to the proceeding; (2) the nature and extent of the petitioner's property, financial, or other interest in the proceeding; and (3) the possible effect of any order which may be entered in the proceeding on the petitioner's interest. An order which permits intervention may be conditioned so as to restrict "irrelevant, duplicative, or repetitive evidence and argument", and in order to have "common interests represented by a spokesmen."[21] This has resulted in consolidated interventions occurring with increasing frequency in AEC proceedings.

A full intervenor may take a position either in support of or against the proposed construction of the plant. It is also possible for interested members of the public to participate by means of a "limited appearance" in order to express their views without becoming a real party to the proceeding.

---

[20] 10 CFR §2.714a.

[21] 10 CFR §2.714d.

The hearings are conducted by a three-man Atomic Safety and Licensing Board,

> one of whom shall be qualified in the conduct of administrative proceedings and two of whom shall have such technical or other qualifications as the Commission deems appropriate to the issues to be decided.[22]

The boards are chosen from a panel of qualified persons appointed by the Commission, a panel which includes nuclear physicists, biologists, health experts, and lawyers. Some are AEC employees; others are selected from the private sector.

Prior to a hearing, board members are supposed to review the application, the ACRS report, the staff safety evaluation, the applicant's environmental report, the Commission's detailed environmental statement, all other papers filed in the cases, and the pertinent rules and regulations of the Commission.[23]

It is worth noting that the board does more than merely weigh the evidence incorporated in the record of the proceeding. The ASLB's were created explicitly to

> facilitate safety determinations and further enhance public confidence by bringing to bear their technical expertise in the resolution of the difficult scientific and technical problems associated with atomic energy licensing.[24]

Clearly the evidence is to be assessed by the Board in terms of their knowledge, experience, and biases.

The ASLB's role under the AEC's Rules of Practice is primarily to determine whether the regulatory staff's review has been adequate to support the

---

[22] 42 USC §2241.

[23] 10 CFR Part 2, Appendix A, §Id.

[24] Green, p. 64.

findings that must be made for issuance of the construction permit. This includes determining whether the applicant has described the proposed design of the facility and identified major safety and health features; whether further technical information which can reasonably be left for later consideration will be supplied in the final safety analysis report; whether safety features and safety research and development programs are incorporated; and whether the proposed facility can be constructed and operated at the selected location without undue risk to the health and safety of the public.[25]

Within 60 days after the notice of hearing has been published, a special prehearing conference is held, in order to permit identification of key issues, take steps necessary for further identification of the issues, consider all intervention petitions to allow preliminary or final determination as to the parties, and to establish a schedule for further actions in the proceeding.[26]

Not until the key issues in controversy are identified in the special prehearing conference order may discovery proceed, and it is limited to these matters. The discovery process is crucial to opposition intervenor groups, because it is really their initial opportunity to actually participate in the evaluation of the applicant's license. As one attorney for an intervenor group put it:

> Intervenors must not only out of preparation, but of necessity, channel the preparation of their case into securing information from unfriendly sources by the only rights they have: that is, the power of subpoena, the right to serve interrogatories and the right to see documents.[27]

----

[25] 10 CFR §2.1046.

[26] 10 CFR Part 2, Appendix A, §II.

[27] Letter from Myron Cherry to ASLB pertaining to Consumers Power Company application to the AEC for a Construction Permit, Docket Numbers 50-329 and 50-330, May 5, 1971.

The AEC Rules of Practice permit parties to an adjudicatory hearing to obtain discovery of evidence by means of written or oral deposition, by written interrogatories, through production of documents or things or permission to enter upon land or other property, and by requests for admission.[28]

Depositions, interrogatories and document production between parties other than the Regulatory Staff are obtainable on notice or request to the other party without involving the ASLB or the Commission. If a deponent or party upon whom a request for production of documents or answers to interrogatories is served fails to respond or objects to the request, the party desiring the information may ask the presiding officer for an order compelling a response or inspection in accordance with the request.[29] The party from whom discovery is sought may seek any order which justice requires to protect a party or person from annoyance, embarrassment, oppression, or undue burden or expense.[30]

Generally, AEC Regulatory Staff documents that are relevant to a proceeding are supposed to be publicly available as a matter of course unless there is compelling justification for their nondisclosure. The Commission's position on discovery of staff documents is that "in view of the comprehensive body of information routinely available without request, there should be minimum need to resort to time consuming discovery procedures."[31]

Discovery against the staff by means of deposition is not permitted, except upon a showing of exceptional circumstances. Interrogatories may be addressed

---

[28] 10 CFR §2.740.

[29] 10 CFR §2.740f.

[30] 10 CFR §2.740c.

[31] 10 CFR Part 2, Appendix A, §IVd.

to the staff only when the information is demonstrated to be necessary to a proper decision in the case and is not obtainable elsewhere.

A party may serve a request for the production of an AEC record or document on the General Manager or Director of Regulation without leave of the Commission or the presiding Board officer. The General Manager or Director of Regulation may object to production on grounds that it is not relevant or necessary to a proper decision in the case, or because the document or information is obtainable from another source, or because it is exempt from disclosure. The requesting party may then apply to the presiding Board officer to compel production of that record or document.[32]

The Freedom of Information Act requires that each agency shall make available for public inspection and copying

> (A) Final opinions . . . as well as orders, made in the adjudication of cases; (B) those statements of policy and interpretation which have been adopted by the agency and are not published in the Federal Register; and (C) administrative staff manuals and instructions to staff that affect a member of the public.[33]

Based on this statutory requirement, the AEC rules provide that

> final AEC records and documents, including but not limited to correspondence to and from the AEC, regarding the issuance, denial, amendment . . . revocation, or violation of a license, permit, or order . . . shall not . . . be exempt from disclosure and will be made available for inspection and copying. . . .[34]

A list of documents exempt from disclosure, based on exceptions contained in the Freedom of Information Act, include records and documents required by Executive order to be kept secret in the interest of national defense or foreign

----

[32] 10 CFR §2.744.

[33] 5 USC §552a.

[34] 50 USC §2.790a.

policy; interagency or intraagency documents containing opinions, advice, or recommendations; proprietary data; and intraagency or interagency memoranda or letters prepared for or by the Commissioners or the Advisory Committee on Reactor Safeguards.[35]

Production of Commission documents is certainly one of the most violently contested aspects of the AEC licensing process by intervenor groups. In hearings before the Joint Committee on Atomic Energy, intervenor advocate Myron Cherry testified that

> The only thing they [the AEC] put in the document room is the application, an ACRS letter, correspondence which says, "Dear Mr. Morris: there is my application." and one from Peter Morris saying, "Dear Mr. Utility: Thanks." They don't put any of the documents in there. I am concerned that every other third memorandum in the Commission gets stamped "confidential" because somebody is stamp happy. The hard facts are not coming out. I can submit to you almost unequivocally that 40% of the time in every licensing hearing has been spent in wrangling over documents which the Commission ultimately released.[36]

Normally within sixty days after discovery has been completed, a second prehearing conference is held to consider simplification, clarification, and specification of the issues; to consider amendments to the pleadings; and to obtain stipulations and admissions of facts as well as the contents and authenticity of documents in order to avoid unnecessary proof. Also a hearing schedule is set and witnesses are identified.[37]

The public hearing is usually held at the nearest suitable place in the vicinity of the proposed plant site. If the hearing is uncontested it may

---

35 Ibid.

36 Hearings before the Joint Committee on Atomic Energy on HR 13731 and HR 13732, 92nd Congress, 2nd Session (March 16, 1972) p. 231.

37 10 CFR Part 2, Appendix A, §IIb.

require as little as a single day's time. In such cases the presiding ASLB's function is to consider whether the application and the record contain adequate information to support the issuance of the construction permit.

However, the majority of construction permit hearings are now contested by one or more intervenor groups. Often these hearings require many weeks of testimony by expert witnesses, depending upon the nature of the matters which are in dispute and the vigor with which opposing intervenors present their case. The hearings are divided into two segments. The first is to consider technical and safety issues involved in construction of the plant; the other is to deal with environmental issues and the adequacy of the Final Environmental Impact Statement.

Every party to a hearing is granted the right to present such oral or documentary evidence and rebuttal evidence and conduct such cross-examination as is required to fully disclose the pertinent facts. The parties are to submit direct testimony of witnesses in writing unless otherwise ordered by the presiding officer on the basis of objections presented.[38]

Normally the applicant makes an oral statement describing the principal safety and environmental issues involved in carrying out the activity sought to be authorized. Other parties to the proceeding also make an oral opening statement describing their position on the proposed licensing action.[39] Unless otherwise ordered by the presiding officer, the applicant has the burden of proof on all issues.[40]

---

[38] 10 CFR §2.743.

[39] 10 CFR, Part 2, Appendix A, §Vc.

[40] 10 CFR §2.732.

The applicant's evidence at the construction permit hearing consists primarily of the amended application, including the preliminary safety report, along with other written testimony. The applicant also makes available a group of witnesses whose oral testimony largely amounts to submission to cross-examination by the staff and any other parties, and by the Board itself, concerning the subject matter of the written evidence.[41]

The staff's safety evaluation, which is adopted orally by staff witnesses, constitutes the core of the evidence on behalf of the staff, supplemented by such oral testimony as may be elicited by the Board.

The proceedings are supposed to be conducted as expeditiously as possible without precluding the development of a clear and adequate record. The order of presenting testimony can be freely varied in the conduct of the hearing; indeed, the Board often finds it helpful to take expert testimony from staff witnesses on a roundtable basis after receiving their prepared testimony into evidence. At any time the Board may interpose its own questions, though it generally allows counsel to complete his examination of witnesses first.

Counsel may object to any questions or line of questioning, or to the admission of any document. The Board may admit the evidence, sustain the objection, or receive the evidence, reserving for a later determination the question of admissibility.[42]

A question may be certified to the Commission or the Atomic Safety and Licensing Appeal Board for determination when a major or novel question of policy, law, or procedure is involved which cannot be resolved except by the Commission

----

41 Kingsley, p. 326.

42 10 CFR Part 2, Appendix A, §5e.

or the Appeal Board and when it is important that the public interest be protected or undue prejudice to a party or significant delay may result.

A party can request the presiding officer to permit a qualified person with scientific or technical training or experience to participate on behalf of that party in the examination and cross-examination of expert witnesses. This is a discretionary decision to be made by the Board, based on a determination as to whether it would further the conduct of the proceeding. If permitted, such examination or cross-examination is limited to areas within the expertise of the person conducting the questioning.[43]

The transcript of testimony and the exhibits, along with all of the papers and requests filed in a proceeding, constitute the record for decision. Generally a decision by the ASLB must be made on the basis of evidence which is in the record; however, the Board is expected to use its expert knowledge and experience in evaluating and drawing conclusions from the evidence in the record.[44]

The Board can request further oral or written argument should it have uncertainties with respect to the matters in controversy. Eventually the presiding officer renders an initial decision. This will constitute the final action of the Commission 45 days after its issuance in any licensing matter, unless exceptions are taken by any party, or the Commission directs that the record be certified to it for final decision.

An initial decision must be in writing, based on the whole record, and supported by reliable, probative, and substantial evidence. It includes (1)

---

[43] 10 CFR §2.733.

[44] 10 CFR Part 2, Appendix A, §5e.

findings, conclusions and rulings on all material issues of fact, law, or discretion presented on the record, along with the reasons or basis for them; (2) all facts officially noticed and relied on in making the decision; (3) the ruling on whether the applicant is granted his license; and (4) the time within which exceptions to the decision, as well as opposition to or support for these exceptions, may be filed.[45]

In addition, at a construction permit proceeding, the ASLB must determine whether the requirements of the National Environmental Policy Act have been complied with, independently consider the final balance among conflicting factors in the record, and decide whether the construction permit should be granted, denied, or appropriately conditioned to protect environmental values.[46]

Within 20 days after service of the initial decision, any party may file exceptions to the decision and a brief in support. Within 10 days after the service of exceptions, any party to the proceeding may file a brief in support of or in opposition to the exceptions. The Regulatory Staff may also file exceptions to the initial decision, or support or opposition to exceptions filed by other parties.[47]

Normally, the review of these exceptions is conducted before a three-man Atomic Safety and Licensing Appeal Board (ASLAB) which performs the review functions otherwise the responsibility of the Commission. In its discretion the Commission itself hears oral arguments upon the request of a party in his exceptions or brief, or upon its own initiative. A party who is dissatisfied

----

[45] 10 CFR §2.760.

[46] 10 CFR Part 2, Appendix A, §Vf.

[47] 10 CFR §2.762.

with the ASLAB's final decision may appeal to the Commission or to the appropriate Federal Court of Appeals.

Within 15 days after a decision by the ASLAB, the Commission also, on its own initiative, may direct that the record be certified to it for review.[48]

Under Commission regulations the AEC may issue the construction permit as soon as the initial decision is handed down. Exceptions to the decision do not interfere with any authorization to issue a construction permit nor do they require that construction be stopped if the permit has been issued. The permit must state the earliest and latest dates for completion of the construction. If the proposed construction of the facility is not completed by the latest completion date, the permit expires unless extended by the Commission. The AEC recognizes developmental problems attributable to the experimental nature of the facility, as well as fire, flood, domestic violence, enemy action, etc. as a basis for extending the completion date.[49]

The AEC prohibits the beginning of construction of nuclear power plants until a construction permit has been issued. This ban covers activities such as clearing of land, excavation, construction of non-nuclear facilities or other substantial action that would adversely affect the natural environment of a site.

However, certain specific activities, such as preconstruction monitoring to establish background information related to the suitability of the site or to the protection of environmental values, is permitted. This includes geologic, seismic, hydrologic, and meteorologic investigations and such clearing and

---

[48] 10 CFR §2.786.

[49] 10 CFR §50.55.

building of roads and physical structures as are reasonably necessary for the purpose of determining site suitability.[50]

Further, Section 50.12 of the AEC regulations provides that

> The Commission may, upon application of any interested person or upon its own initiative, grant such exemptions from the requirements of the regulations in this part as it determines are authorized by law and will not endanger life or property or the common defense and security and are otherwise in the public interest.[51]

This section has been used to permit the performance of extensive preparatory construction activities, including the placing of concrete for plant structures prior to the issuance of a construction permit.[52]

After about two years of construction work, a final technical safety analysis (FSAR) is submitted by the applicant to the Commission. The FSAR must include information that describes the facility, presents the design bases and the limits on its operation, and prevents a safety analysis of the structures, systems, and components, and of the facility as a whole.[53]

The applicant must describe the equipment and procedures installed for the control of gaseous and liquid effluents and for the maintenance and use of equipment installed in radioactive waste systems. In addition he must also submit a revised estimate of the quanity of expected releases of radio-nuclides, if the expected releases and exposures differ significantly from the estimates

---

50 Step-by-Step Procedure In Licensing of Nuclear Power Reactors, AEC Office of the Director of Regulation (1972) p. 3.

51 10 CFR §50.12.

52 Letter from Harold Price (AEC) to Consumers Power Company, July 30, 1970.

53 10 CFR §50.34.

submitted in the application for a construction permit.[54] In addition, the applicant must submit technical specifications on expected effluents from the reactor.[55]

The ACRS again reviews the project and furnishes its advice to the Commission in the form of another letter which lists problem areas that require further work by the applicant. After favorable completion of review by the Regulatory Staff, the Commission must publish a notice of intent to issue an operating license to applicant, giving at least 30 days advance notice. The notice presents the position of the AEC Regulatory Staff and the ACRS.[56]

In the meantime the applicant files an Environmental Report in conjunction with its operating license application. This report is an updating of the applicant's "Environmental Report--Construction Permit Stage", and is supposed to (1) discuss differences between currently projected environmental effects of the nuclear power plant and the effects discussed in the Environmental Report submitted at the construction stage; (2) discuss the results of all studies which were not completed at the time of pre-construction review, and indicate how these results were factored into the design and proposed operation of the plant; (3) describe the monitoring programs which have been and will be undertaken to determine the effects of the operating plant on the environment; and (4) discuss any planned studies that may yield results relevant to the environmental impact of the plant.[57]

----

[54] 10 CFR §50.34a.

[55] 10 CFR §50.36.

[56] Step-by-Step Procedure, p. 4.

[57] "Guide to the Preparation of Environmental Reports", pp. 2-3.

The Commission receives this second Environmental Report from the applicant and develops a Final Environmental Statement according to the same procedure it used to produce a final statement at the construction permit stage. Subsequent hearings and action on the environmental aspects involved in issuance of an operating license are based on this statement.

Within 30 days after the AEC's notice of intent to issue an operating license is published, any person whose interest may be affected by the proceeding may petition the AEC to hold a hearing. A hearing is not mandatory at the operating license stage; if no one requests it, the AEC can issue an operating license without a hearing.

Should a request be received for a hearing and should it be granted, the hearing proceeds in much the same fashion as described for the construction permit stage. In an operating license proceeding, however, the Board will determine only the matters in controversy among the parties as the issues to be decided, within the purview of certain types of considerations. These include: (1) whether the facility will operate in conformity with the application as amended, the provisions of the Act, and the rules and regulations of the Commission; (2) whether issuance of the license will be inimical to the health and safety of the public; and (3) whether, with respect to requirements of §102(2)(C) and (D) of the National Environmental Policy Act, the operating license shoudl be issued as proposed.[58]

The presiding ASLB, which is generally composed of different individuals than those who sat at the construction permit hearings, makes findings only on

----

[58] 10 CFR Part 2, Appendix A, §VIIb.

the matters in controversy. Depending on how the Board resolves these matters, the Director of Regulation issues, denies, or appropriately conditions the operating license.[59]

Typically the ASLB's decision authorizes a maximum power level at which the facility can be operated in accordance with the technical specifications reflected in the evidence in the proceeding, and specifies a term of years for the license, not to exceed 50. Other conditions are included which are intended to reflect the Commission's desire to protect the public health and safety as well as to consider environmental values.

A recent amendment to the Atomic Energy Act of 1954 (PL 92-307, 1972) permits the Commission to permit temporary operation of a nuclear reactor at up to full power, for a period of six months, before the Commission's study of the environmental impact is completed.[60] A public hearing must be held before the temporary license is granted; however, the hearing is to be conducted with such

> expedited procedures as the Commission may . . . deem appropriate for a full disclosure of material facts on all substantial issues raised in connection with the proposed temporary operating license.[61]

The findings which must be made by the ASLB before granting the temporary license fall into three categories: (1) that construction has been completed and provisions relative to protecting the health and safety of the public have been met; (2) that temporary operation provides adequate protection of the environment; and (3) that granting of the temporary license

---

[59] _Ibid._

[60] 42 USC §2242.

[61] 42 USC §2242(a).

> is essential toward insuring that the power generating capacity of a utility system or power pool is at, or is restored to, the levels required to assure the adequacy and reliability of the power supply, taking into consideration factors which include, but need not be limited to, alternative available sources of supply, historical reserve requirements for the systems involved to function reliably, the possible endangerment to the public health and safety in the event of power shortages and data from appropriate Federal and State governmental bodies. . . .[62]

The action of the presiding officer at a temporary operating license hearing is taken without prejudice to the position of any party to the proceeding for the issuance of the full-term operating license.[63]

The appeal process in the event exceptions are filed to an initial decision granting a regular operating license stage is the same as the procedure at the construction permit stage. Once an operating license is issued, the plant is monitored by the Directorate of Regulatory Operations to insure that the applicant complies with AEC regulations and the license specifications.

RULE MAKING

Like other administrative agencies, the AEC sometimes tries to resolve questions of policy by means of legislative or rule making hearings. In the last year or so this has occurred whenever a generic issue has been raised by intervenor groups at several licensing proceedings, and the AEC has determined that it would be more expeditious to hold a single hearing or set of hearings

---

62 42 USC §2242(b). In granting the only temporary operating license that has been issued, the Commission determined that this statute does not require inquiry into historical reserve requirements as a condition precedent to its granting of such a license. See ASLAB memorandum in the Vermont Yankee Nuclear Power Proceeding, Docket Number 50-271, September 5, 1972.

63 10 CFR §2.606.

in order to make a final policy determination. Rule making hearings have been used by the AEC to consider permissible levels of radioactive releases, criteria for Emergency Core Cooling Systems (ECCS), and matters relating to the nuclear fuel cycle.

The Commission has established rules of practice that govern rule making proceedings.[64] According to these rules, "rule making may be initiated by the Commission at its own instance, on the recommendation of another agency of the U.S., or on the petition of any other interested person."[65] Any interested party may also petition the Commission to "issue, amend, or rescind any regulation."[66]

The Commission affords interested persons an opportunity to participate in rule making through the submission of statements, information, opinions, and arguments in the manner stated in the notice of hearing.[67] The notice of hearing is also to include "the terms or substance of the proposed rule, or a specification of the subjects and issues involved."[68] Whenever the Commission adopts a rule or regulation, it issues a general statement of its basis and purpose, and publishes the notice and regulation in the Federal Register.[69]

It should be pointed out that an agency can utilize both formal and informal rule-making proceedings. Formal hearings are open to the public and permit

----

64 10 CFR Part 2, Subpart H.

65 10 CFR §2.801.

66 10 CFR §2.802.

67 10 CFR §2.805.

68 10 CFR §2.804.

69 10 CFR §2.806.

oral testimony by all interested parties, while informal hearings generally consist of the agency accepting written comments from interested parties without providing public hearings. However, even formal rule-making hearings differ significantly from licensing hearings in that they use a "legislative-type" format; that is, a format in which each party presents his testimony without subjecting himself to depositions, written interrogatories or questioning by parties other than the hearing board. An exception was made in the ECCS hearings, where an experimental quasi-judicial procedure was adopted which allowed cross-examination of witnesses and extensive exchanges of documentary evidence among the various parties and the AEC. The detailed rules of that proceeding evolved through the course of the hearing; decisions were made by the hearing board and the Commission throughout on an *ad hoc* basis in order to resolve the numerous unforeseen procedural issues which arose in this quasi-judicial, hybrid-type hearing.

CHAPTER III: THE CASES

MIDLAND PLANT UNITS 1 AND 2

The proposed Midland Plant (Units 1 and 2) is to be located in Midland, Michigan on the south shore of the Tittabawassee River. The two pressurized-water nuclear reactor systems will generate 1300 megawatts of electricity for use by Consumers Power Company of Michigan and the Michigan Power Pool as well as 4,050,000 pounds per hour of steam which will be used exclusively by the Dow Chemical Company's Midland plant for chemical processes and heating.

The location of the plant was dictated, in large measure, by the proximity of the Dow Chemical plant. The site is flat land at the edge of the city of Midland. The Dow Chemical Company's complex is immediately across the river. The land surrounding the site is partly open farmland and partly suburban residential. The population within one mile of the plant consists of 38 residents (and a business-loading population of 2,145); within five miles the resident population is 41,000 (with a business-loading population of 34,000); and within 50 miles the projected 1980 population is about 1,100,000. The Bullock Creek Elementary School (enrollment: 493) is slightly over one mile from the proposed plant structures.

An 880-acre man-made cooling pond will be used to dissipate most of the waste heat to the atmosphere. A supplementary cooling system will limit the temperature of cooling pond blowdown to the river to within one degree Fahrenheit of the river water temperature.

On January 13, 1969, Consumers Power Company (CPC) of Jackson, Michigan filed with the AEC an application for authorization to construct and operate the plant at the Midland site. In a letter to Peter Morris of the AEC's Division of Reactor Licensing, Robert Allen, Senior Vice President of CPC noted that:

> It was our understanding that, by having submitted our application as we did informally in November 1968, and by having reviewed at an early date with the ACRS the site-related features of the Midland Plant, we could expect to achieve a construction permit date of March 1, 1970.[1]

By September 1969, the CPC schedule was more modest calling for public hearings on the license by mid-April of 1970.[2] Communication between the AEC staff and CPC continued into 1970. On June 18, 1970, the Advisory Committee on Reactor Safeguards reported to AEC Chairman Seaborg that it had completed its review of the Plant. Noting that "Midland is the first dual purpose reactor plant to be licensed for construction" the ACRS discussed several aspects of the plant including certain safety-related items and concluded:

> The Committee believes that the above items can be resolved during construction and that, if due consideration is given to those items, the nuclear units proposed for the Midland Plant can be constructed with reasonable assurance that they can be operated without undue risk to the health and safety of the public.[3]

The public docket on the Midland plant indicates that the first public reaction to the plant came from Wendall Marshall of Mapleton, Michigan in July, 1970 in the form of a telegram to Chairman Seaborg requesting information concerning the cooling pond and its possible fogging effect.[4]

In May, 1970, CPC had requested an exception which would permit certain construction work at the Midland site to proceed prior to the issuance of the

----

[1] Letter from Robert Allen, Consumers Power Company, September 11, 1969.

[2] Ibid.

[3] Letter from Joseph M. Hendrie, Advisory Committee on Reactor Safeguards, June 18, 1970.

[4] Telegram from Wendell Marshall, July 21, 1970.

construction permit. On July 30, 1970 the AEC replied that

> We have reviewed your request for an exemption . . . and have concluded that the granting of the exemption is authorized by law, will not endanger life or property or the common defense and security and is otherwise in the public interest. Accordingly, the Commission hereby authorizes you, at this time, to proceed with the placement of the concrete. . . . I wish to emphasize that the granting of this exemption shall have no bearing upon the subsequent granting or denial of a construction permit. . . .[5]

CPC's environmental statement was submitted on July 24, 1970 and made available for comments from state and local agencies.

In August, 1970 citizen interest in the Midland Plant began to grow. At a hearing held by the staff of the Michigan Water Resources Commission, Mary Sinclair of Midland presented testimony which pointed out the "AEC's questionable attitude toward public health and safety." Quoting from various scientists, she outlined controversies on several aspects of nuclear power plant operation, and concluded:

> I would like to reemphasize that the serious disputes among scientists, central to public health and safety, on the effects and disposal of radioactive wastes make it vitally important that no such wastes are released to our waterways. They are long-lived and their damage would be irreversible. The public cannot allow this miscalculation both for our time and for our future generations.[6]

In October, 1970, the AEC named an Atomic Safety and Licensing Board to conduct hearings on the construction permit. The Board consisted of Clark Goodman, David Hall, and Arthur Murphy.[7] The notice of hearings stated that:

---

[5] Letter from Harold Price, AEC, July 30, 1970.

[6] Testimony of Mary Sinclair before Michigan Water Resources Commission, August 10, 1970.

[7] Mr. Murphy is professor of law at Columbia University School of Law; Dr. Goodman is an engineer, and professor of physics at University of Houston; Dr. Hall is a physicist employed by the Los Alamos Scientific Laboratory.

> . . . the board will, without conducting a de novo evaluation of the application, consider the issues of whether the application and the record of the proceeding contain sufficient information, and the review of the Commission's regulatory staff has been adequate, to support the findings proposed to be made and the construction permits proposed to be issued by the Director of Regulation.[8]

Petitions to intervene in the proceedings were filed by several parties. One of the intervenors was the Midland Nuclear Power Committee, a group of about 50 people representing "trades, professions, businesses, organizations, and industries." The Committee, chaired by Dr. Wayne North, Senior Minister of the First Methodist Church of Midland, concluded its petition by noting that:

> . . . it is in its best interest and in the best interest of all citizens of the area surrounding the proposed plants that the Board render a favorable decision as to this application, and grant licenses for construction of Midland Plant Units 1 and 2 to the Consumers Power Company.[9]

The Dow Chemical Company also petitioned to intervene

> . . . in view of its intimate continuing involvement with the health and safety, as well as economic well being, of the Midland, Michigan area and in light of its announced contract to purchase from the proposed nuclear power plant its requirements of process steam and electrical energy for its Midland Chemical plant complex.[10]

The Dow petition continued:

> Dow is interested in this proceeding because it desires to assure that the construction permit will issue and be in such form as to assure that the stringent requirements for the quality of Dow's process steam will be met.[11]

----

8 "Notice of Hearing on Application for Construction Permits", October 27, 1970.

9 Petition for Intervention, November 10, 1970, p. 5.

10 Petition by the Dow Chemical Company for Leave to Intervene, November 11, 1970, p. 1.

11 Ibid., p. 3.

The Environmental Defense Fund (EDF), a non-profit, tax-exempt membership corporation made up of "scientists and other citizens dedicated to the protection of man's environment", also filed a petition to intervene noting that

> Part of the mission of EDF is to guarantee that whenever major federal actions which may affect the environment are proposed a full exploration of the impact of the project on the environment and alternatives to it is conducted.[12]

A group of citizen's organizations (hereafter known as the Saginaw Intervenors)[13] filed a common petition to intervene in the construction permit hearings. The group included: the Saginaw Valley Nuclear Study Group, an

> association comprised of citizens and residents of Michigan . . . all of whom are interested in the dissemination of information and stimulation of public awareness and involvement in the study of nuclear power

which was concerned

> . . . that the construction and operation of the Midland Plants will result in an unsafe and inefficient exercise in the use of atomic energy, will create deleterious and adverse effects upon the City of Midland, Michigan and elsewhere, will contribute to and create adverse environmental effects which can be avoided. . . . In addition . . . the Saginaw Valley Nuclear Study Group . . . [is] concerned about the imposition of a nuclear plant, or plants, essentially situated within the heart of a small community;[14]

the Citizens Committee for the Environmental Protection of Michigan, a group of citizens concerned

---

[12] Petition for Leave to Intervene, November 12, 1970, p. 3.

[13] Saginaw Valley, et. al., Intervenors, include the following intervenors: Saginaw Valley Nuclear Study Group, Citizens Committee for Environmental Protection of Michigan, Sierra Club, United Auto Workers of America, West Michigan Environmental Action Council, and University of Michigan Environmental Law Society.

[14] Petition to Intervene, p. 2.

> . . . that their life styles, their occupations, and their investments, both emotional and financial, will be threatened by the construction and operation of the proposed plants. . . . [They] are also concerned because they live geographically close enough to the proposed Midland Plants to be seriously and adversely affected by the radioactive and gaseous effluents which are planned to be emitted from the Midland Plants . . . ;[15]

the Sierra Club, a national conservation organization representing

> . . . the public interest in protecting the natural resources of the nation and . . . opposed to the exploitation of such resources contrary to environmental enactments of the federal and local governments;[16]

the United Auto Workers of America, a labor organization with concerns similar to those expressed by the other petitioners as well as a concern that

> . . . members of the UAW throughout the United States may be adversely affected by virtue of the fact that the steam proposed to be generated by the Proposed Plants may adversely and deleteriously affect products produced by companies who presently plan to use or may use the steam generated from such Proposed Plants;[17]

Trout Unlimited, a membership organization concerned, among other things, "with opportunities for quality fishing by maintaining and ameliorating trout fishing lakes, rivers and streams such as the Tittabawassee River;"[18] West Michigan Environmental Action Council, Inc., concerned with "nuclear safety, radiological protection and the many broad environmental questions which must be considered . . . ;"[19] the University of Michigan Environmental Law Society, a group of law students which wished

---

15 *Ibid.*, p. 3.

16 *Ibid.*, p. 4.

17 *Ibid.*, p. 6.

18 *Ibid.*, p. 7.

19 *Ibid.*, p. 8

> . . . to demonstrate to the people of the United States and Michigan, particularly the young activist community, that environmental protection can be achieved through established administrative and legal channels. Accordingly, such Law Students opt for intervention to make it plain that large financial or ingrained government interests may not violate any law without being called into account.[20]

The petition lists numerous contentions in the areas of nuclear safety, financial qualifications of the applicant, technical qualifications of the applicant and the construction team, standards for protection against radiation, environmental effects, and the role of the Atomic Energy Commission.

A group of individuals residing at or near the village of Mapleton and within close proximity of the proposed plant launched a separate intervention. According to their petition to intervene,

> they are interested in the protection of the air, water, and other natural resources and the public trust therein from pollution, impairment, or destruction. . . . [S]hould the Honorable Commission grant leave to actively intervene and participate in this matter, they intend to have an independent safety analysis made of theimpact on the environment by a qualified expert in thermal and atmospheric polution [sic].[21]

The Mapleton Intervenors further asserted that

> the AEC has failed to submit a detailed environmental statement as to the probable effects on the environment . . . and that the Applicant's proceedings before the Safety and Licensing Board are premature or illegal. . . . They [the Intervenors] are further interested in the protection of vested property rights . . . and/or prevention of any possible or probable hazzards [sic] or dangers to human health by radioactive and gaseous effluents which are planned to be emitted from the Midland Plants during normal operation; and/or possible or probable detrimental or deleterious radiological and thermal effects, as well as the results of wastes and other effluents

---

[20] _Ibid_., p. 8.

[21] _Petition for Leave to Intervene_, November 28, 1970, p. 1.

> ultimately discharged into the navigable Tittabawassee River; and/or possible or probable adverse atmospheric conditions resulting from the operation of said plants, including, but not limited to fog. . . .[22]

The Mapleton petition concluded in a rather unique manner, stating that

> Based on the foregoing, your Petitioners most respectfully pray that they be granted leave to intervene and actively participate in this matter.
>
> AND YOUR PETITIONERS WILL FOREVER PRAY.[23]

During the course of the hearings, the various intervenors sought to raise various issues for consideration. The Midland Nuclear Power Committee participated as an intervenor in complete support of the Midland Plant and sought to demonstrate the safe design of the plant, the environmental advantages of nuclear power over alternative means for producing electricity, and the need for the facility to provide adequate competitive energy services to the state of Michigan and the Midland area. Dow Chemical Company raised certain legal and procedural questions including the following:

1. Are the environmental considerations set forth by the opposing intervenors properly before the ASLB in light of NEPA? Can they raise an unlimited number of issues?

2. Has the public's right to an early determination of the issues been completely denied by the "delaying tactics" of the intervenors?

3. Should the other commitments of Saginaw Intervenors' Attorney Cherry be permitted to interfere with the orderly process of litigation?

---

22 *Ibid.*, p. 3.

23 *Ibid.*, p. 4.

Among the major safety and technological issues which the Environmental Defense Fund sought to have considered in the hearing were:

1. Is there presently sufficient knowledge or experience with pressurized water reactors of the size and type of proposed Midland Plants to justify anything but the most conservative approach to siting of the reactors?

2. Is there reasonable assurance that the emergency core cooling system can function as quickly as necessary or with the reliability that is demanded if meltdown is to be avoided? (Because ECCS effectiveness was challenged in numerous licensing hearings, the AEC has consolidated the ECCS question into a single rule-making proceeding held in Bethesda, Maryland.)

3. Does the maximum hypothetical accident (MHA) take into consideration the generation of large quantities of hydrogen gas within the containment after a loss-of-coolant-accident?

4. Does the MHA take into account the synergistic and counter-productive effects of various systems used to control the release of iodine?

5. Have adequate emergency plans and procedures been developed with respect to an accident which would require immediate evacuation?

One of the primary reasons for EDF's participation was to assure compliance with the National Environmental Policy Act. Anthony Roisman, the attorney for EDF, commented at the start of the hearing:

> We do not believe that a hearing can be conducted into the question of whether a construction permit can be issued until the environmental statements required by NEPA have been prepared, submitted to the parties for consideration, so that the hearing may factually look into matters relating to the environment.[24]

---

[24] Transcript, p. 14.

EDF sought to raise a number of environmental issues such as:

1. Have applicant or the AEC radiation standards taken into account dangerous illnesses or injuries, if any, which may occur from radioactive effluents reaching man and combining with chemical effluents from Dow?

2. Has there been a valid analysis of the various alternatives to using nuclear sources of generating electricity, as required by NEPA?

3. What effect will the proposed plants have on the recreational activities currently enjoyed within the area of immediate environmental impact?

4. How will the operation of the plants affect the water quality of the Tittabawassee River at a time when community leaders are committed to upgrading the river?

5. Has adequate consideration been given to the evaporation and fogging problems and other atmospheric effects occurring as a result of the daily operation of the cooling ponds?

6. Should the AEC's NEPA statement discuss the procedures and facilities for the handling, storage and shipment of spent fuel elements?

Finally, EDF raised certain legal and procedural issues:

1. Should the Commission consider whether its objectivity is colored by the fact that its promotional objectives are inconsistent with its regulatory objectives?

2. Should the Commission supply the EDF with any documents used, relied upon or prepared by the Regulatory Staff or any other office of the AEC, including the Preliminary Safety Evaluation Report?

3. Has NEPA implementation been violated by Applicant's submission of an inadequate and non-complying report, and the Commission's failure to make the

type of environmental study required by the Council on Environmental Quality (CEQ) regulations?

The Saginaw Intervenors pursued many of the issues raised by EDF. One question they sought to have answered could be put as follows: Does the real difficulty with analyzing the propriety of building a nuclear reactor lie in the overwhelming commitment made by the applicant before anyone has an opportunity to begin the analysis? They also raised certain environmental and procedural questions:

1. Do dry cooling towers represent a better alternative than the proposed cooling pond?

2. Does the proximity of the plant to major population centers in Michigan seriously threaten the health and welfare of the local residents?

3. Have the costs involved in building the plants been given equal consideration with the benefits?

4. Have all possible alternatives been examined by the AEC as required by Section 102 2(c) and (d) of NEPA?

5. Should not the public interest intervenors in this application be regarded as legal defendants, thereby transferring the burden of proof to the applicant to prove that the environmental considerations raised by the intervenors are invalid?

The Mapleton Intervenors raised similar questions as well as a set of environmental issues:

1. Is the meteorological analysis of the proposed site sufficient?

2. Are the environmental effects from a fossil fuel plant less severe and long lasting than those from a comparative nuclear plant?

3. Will the proposed plants degrade state and local air quality?

4. Will the proposed plants allow Dow Chemical to close down inefficient fossil-fired plants, and also enable applicant to operate its older, less efficient units a smaller percentage of the time?

The first prehearing conference was held on November 17, 1970. After that conference there was considerable communication between the Board and attorneys for the various parties over the availability of documents. The Board had ordered that all parties should submit a list of witnesses and a summary of their testimony. Myron Cherry, attorney for the Saginaw Intervenors responded that

> The only documents in the possession of these intervenors at this moment which seem relevant are the Applicant's Preliminary Safety Analysis Report and the Staff Safety Evaluation Report. This should come as no surprise to anyone since intervenors have not yet had an opportunity to examine the many documents in the possession of the Applicant, the Staff, Dow Chemical Company and others which are relevant and which may lead to the discovery of other relevant documents by intervenors.
>
> . . . .
>
> Intervenors cannot at this point state who their witnesses may be. . . . It is important to remind the Board and the parties that the intervenors contesting the issuance of a license are in a real sense defendants in this proceeding and it is the Applicant, the staff and the supporting intervenors who must carry the day on their burden of proof.[25]

Mr. Cherry also claimed that the unavailability of the AEC's environmental impact statement, the uncertainty as to AEC jurisdiction or responsibility for thermal effects (pending the outcome of the Calvert Cliffs case), and the proprietary nature of some of the documents in the case made his preparation of a

---

[25] Letter from Myron Cherry to ASLB, December 15, 1970.

case difficult if not impossible. Robert Lowenstein, representing CPC, criticized Mr. Cherry's "plainly dilatory tactics"[26] and urged that AEC move ahead with the proceeding. Milton Wessel, representing Dow, agreed noting:

> The public is the key party to this proceeding, and is anxiously awaiting its outcome. Yet to this point its right to an early determination--one way or the other--has been completely denied by the Saginaw intervenors' refusal to obey either the letter or the spirit of the Board's directions. . . . Why does the Board listen only to the Saginaw intervenors? Why doesn't it tell those supporting the application what it is interested in on behalf of the public, so they can prepare and present evidence. If the Saginaw intervenors wish to participate, fine; if not, the matter should proceed just as it would have had they not chosen to participate. . . . How long must we--and especially the public--continue to accept and submit to the Saginaw intervenors' decision that there shall be no Midland Plant?[27]

An ASLB order in March, 1971 attempted to set a schedule and program for the orderly conduct of the proceeding including the setting of dates for the filing of interrogatories and direct evidence. The Board noted that:

> The central problems of organizing this proceeding seem to revolve about the nature of the role of intervenors. . . . As all are aware, the Board has granted all requests for intervention including those which were technically untimely and those as to which the demonstration of interest is less than crystal clear. In doing so, the Board feels it acted consistently with the policies of the Atomic Energy Commission to encourage public participation, and also with the clear trend of recent court decisions.
>
> . . . .
>
> At least as to radiological hazards, the role of the Board in this proceeding is, in its essence, to make findings as to the safety of a reactor of the general design proposed in the particular locality. The difficulties of performing that role in the context of a public hearing are not inconsiderable. . . . [T]he primary function of the intervenor is to assist the Board in making its safety evaluation.[28]

----

26 Letter from Robert Lowenstein to Arthur Murphy, February 23, 1971.

27 Letter from Milton Wessel to Arthur Murphy, February 25, 1971.

28 Order, March 3, 1971, pp. 3-4.

In the order the Board took issue with the contention that the role of intervenors could be likened to that of defendants in a criminal trial; "insofar as they [the intervenors] claim to be representing the public interest, they have voluntarily accepted a burden of representation of the public interest."[29] Speaking for the Board, Mr. Murphy continued:

> . . . [T]he Board must pass on very complex matters. Over two years have gone by since the filing of the application. . . . A mountain of documents has accumulated. It seems almost self-evident that competent counsel, not under any compulsion to limit the time devoted to the case, can, by utilizing the normal litigation techniques . . . prolong the hearing almost indefinitely. In our view such prolongation would not be in the public interest.
> . . . .
> Although we believe that as a general rule a litigated proceeding must be conducted by counsel, we believe that counsel, unsupported by technically qualified personnel, are unlikely to make a significant contribution to safety.[30]

Throughout early 1971, the AEC received numerous letters from citizens and even Congressmen protesting delays in the granting of a construction permit.

The intervenors, particularly EDF and the Saginaw intervenors, continued to maintain that in the absence of the necessary studies and analysis it was impossible to identify specific impacts of the plant. Anthony Roisman (EDF attorney) claimed in a letter to the Board that it was not realistic to blame delays on the intervenors when, 15 months after the enactment of NEPA, the AEC and the applicant were unprepared with respect to environmental issues.[31]

The debate over delays, production of documents, and filing of interrogatories continued through the spring of 1971 (and in fact, continued, almost

---

29 Ibid., p. 5.

30 Ibid., p. 6.

31 Letter from Anthony Roisman to Arthur Murphy, March 11, 1971.

without respite, throughout the proceeding). Letters and briefs of explanation and accusation among the counsel for CPC, Dow, and the intervenors fill the voluminous public record of the proceeding. As the Calvert Cliffs case worked its way through the courts, the debate expanded to include questions of the responsibility of the AEC to consider non-radiological environmental issues. Some highlights of the record are included in this brief description of the Midland case; readers interested in greater detail are referred to the public record of the case:

> Mr. Cherry: (letter to Chairman Murphy, May 5, 1971) I am sure the Board is aware that opponents of a construction permit for a nuclear plant cannot easily find witnesses or indeed assistance from the ranks of industry or academicians who for one reason or another are so geared to pursuing their professions that they cannot render objective assistance to intervenors. Accordingly, intervenors must, not only out of preparation, but of necessity, channel the preparation of their case into securing information from unfriendly sources. . . . For this Board to render decisions which destroy intervenors' rights with respect to adequate pretrial preparation by way of inspection of documents and interrogatories is, in the light of the circumstances, tantamount to denying the intervenors the right to prepare and hence participate.
>
> The important and far-reaching issues in this proceeding are not whether a plant in Midland, Michigan gets built, but whether or not administrative agencies, by virtue of their proceedings engender respect in the legal community and indeed with the public at large. An ill-prepared intervenor does little to further the cause of public hearings, and a Board who does not bend-over backwards to help an intervenor prepare, does little to follow the mandate of Congress to hold a public hearing upon the issue of a construction permit for a nuclear power plant.[32]
>
> Mr. Roisman: (letter to Chairman Murphy, May 17, 1971) . . . [W]e cannot share the view of some parties to this proceeding which recklessly use the terms "unreasonable delay" as though

---

32 Letter from Myron Cherry to Arthur Murphy, May 5, 1971.

> the mere utterance of the terms provides an answer to any and all requests for discovery. The suggestion that public interest intervenors should become familiar with the Midland nuclear plant in a few months when the Staff and applicant have had several years of familiarity with nuclear plants in general and this plant in particular is absurd. . . . It is axiomatic that so long as the AEC persists in allowing the public to intervene in proceedings only after the applicant and the Staff have completed their review and are in fact ready to issue a construction permit and begin construction that delay will occur. But it is unconscionable for responsible parties who know the complexities of these matters to claim that such a delay is unreasonable.[33]

> Chairman Murphy: (Order of May 19, 1971) Turning to the question of whether the documents, although within the class of internal working papers, should be disclosed, the Board has found the decision difficult. "Need" and "relevance" and "public interest" are elastic concepts. In balancing the various considerations, the Board has been mindful that the proliferation of interventions in licensing proceedings is a reflection of frustration on the part of many members of the public that the public voice is being inadequately considered. However ill-founded that feeling may be, it seems clear that denial of access to documents, except for good reason, will only enhance the frustration. Accordingly, the Board should not, in our view construe the requirement of need or relevance too narrowly.[34]

In late spring (1971) questions about the adequacy of performance of emergency core cooling systems were raised by Mr. Cherry. Until the generic rule making hearings on ECCS began in Bethesda in early 1972, the question of ECCS became a major point of contention and the debate continued:

> Mr. Lowenstein: (attorney for CPC) Saginaw Intervenors seem to be seizing at random upon every pretext which comes to hand, first to delay discovery and now, at the eve of the hearing to extend discovery.[35]

---

[33] Letter from Anthony Roisman to Arthur Murphy, May 17, 1971.

[34] Order with Respect to Documents as to Which Privilege is Claimed by AEC Staff, May 19, 1971, pp. 2-3.

[35] Letter from Robert Lowenstein to ASLB, May 20, 1971.

Prior to hearings in July of 1971, Mr. Cherry submitted a fairly detailed list of topics of cross-examination on behalf of the Saginaw Intervenors. However, he noted that:

> The Board should be mindful that a substantial portion of the Intervenors' case will be to probe and demonstrate that sufficient uncertainty exists in Applicant's presentation so that it cannot be assumed that the proposed Units as a total engineered system satisfies and supports a finding of reasonable assurance of no undue risk.[36]

Meanwhile the debate continued:

> Mr. Lowenstein: In addition to the information contained in the voluminous responses by applicant to Saginaw intervenors' interrogatories, applicant made available to Saginaw intervenors substantially in excess of 15,000 pages of documents for inspection. . . . Not withstanding the time--approximately six months--heretofore available for discovery purposes and the vast amounts of information made available to them, the Saginaw and Mapleton intervenors have failed to specify a single specific respect (and grounds therefore) in which the application fails to comply with AEC regulations or in which the plant site or design is inadequate to protect the public health and safety.[37]
>
> Mr. Cherry: The Board has not permitted intervenors to rely upon or use the claimed proprietary information of a vendor other than B&W [Babcock and Wilcox], and particularly Westinghouse. . . .[38]

The Calvert Cliffs decision of July, 1971 opened the proceeding to non-radiological issues which prolonged the proceeding still further.

Meanwhile letters to Congressmen, the President, and the AEC demonstrated considerable public annoyance at the delay in arriving at a decision. The excerpts below are typical of the correspondence:

---

36 Letter from Myron Cherry to Arthur Murphy, June 10, 1971.

37 Letter from Robert Lowenstein to Arthur Murphy, June 19, 1971.

38 Intervenors Suggestions as to Questions to be Certified with Respect to the Use of Comparative and/or Proprietary Information, July 12, 1971, p. 1.

> Dear President Nixon:
>
> Would you please get the AEC on the ball. Consumers Power Co. has been trying for 4 years to get a permit to build a Nuclear Power Plant here . . . so they can sell power and steam to Dow Chemical Co. . . . I work at Dow and I am worried about my job. There's a small group (10 or 15) people fighting this Nuclear Plant they have lots of money and are using it on lawyers to stop this plant. We need this plant now. We need jobs. We know its safe because they are all over the country and in submarines.[39]
>
> . . . .
>
> Dear Congressman:
>
> We are very concerned regarding the present status of the proposed Nuclear Plant for the Midland, Michigan area. We have confidence that Consumers Power is a responsible company and has designed the proposed plant with the latest technology to insure its safety. However, in spite of the confidence of the majority of people in Consumers and the stated safety of the proposed plant, a very few people have been able to stall the issuance of a construction permit for--years. . . . We would like to see some initiative from those elected, to dig in, get the facts. . . . [Y]ou may just come to the conclusion that the plant must be built and take its fate out of the bureaucratic red tape.[40]

The Midland Daily News expressed its opinion and that of the "Midland Nuclear Power Committee and Others" in a front page letter on August 6, 1971 entitled "Open Letter to the Public: Will a Few People Destroy Our Area" which said in part:

> The economic growth and future of Midland received another setback by last week's court ruling in Maryland which says that the Atomic Energy Commission must also question environmental issues.
>
> . . . .

---

39 Letter from Walter H. Swiger, Jr., August 18, 1971.

40 Letter from Mr. and Mrs. Duane Dosson to Representative James Harvey, August 17, 1971.

> We believe . . . that the intervenors have played a role in Consumers Power Company's careful reappraisal of all aspects of the . . . Plant which has resulted in some alterations--some of which are very speculative as to the real need. The changes and the delays have had a staggering effect on the cost of the project. What started out two years ago estimated at $267,000,000 is now approaching $500,000,000. That in itself is a sobering thought.
>
> . . . .
>
> The intervenors, only a handful of whom can be identified as local people, seemed bent on saving us from something--just what though we are not sure. We are not experts, nor have we professed to be, but we have listened to, discussed with and paid attention to those who are true experts in nuclear power and are satisfied that the majority of real experts have done their homework well. . . . The Midland Nuclear Power Committee has put several questions to the "Study Group" intervenors. Questions such as who are the members of the Saginaw Valley Nuclear Study Group and where do they live? Where are the funds coming from to retain Myron Cherry and pay their other expenses? What changes would satisfy the intervenors? What do they propose to replace the jobs and tax base that will be lost if Dow Midland Division phases out?
>
> . . . .
>
> We believe that those few who want to live in the past should know that the vast majority of the area people want to move into the last quarter of the twentieth century! They should know that we are not afraid of the new, but welcome each new step of progress![41]

Public acceptance of nuclear power in Midland was clearly widespread. The Midland County Board of Commissioners voted unanimously to accept a resolution "that sets as the board's goal final completion of an atomic energy plant in Midland"[42] and to appropriate $20,000 to accomplish the goal. Commissioner Lynn O. Francis is quoted as criticizing two opponents of the plant, "saying he can't understand why Wendell Marshall and Mrs. William D. Sinclair can bring this area to virtually a standstill."[43]

---

[41] Midland Daily News, August 6, 1971, p. 1.

[42] Ibid., August 17, 1971.

[43] Ibid.

Some area newspapers were considerably more skeptical about the need for and advisability of building a nuclear plant. Commenting on the Board of Commissioners' promotional bent, the Saginaw News noted:

> A level of absurdity has been reached in the welter of action and reaction springing from quasi-legal argument and seemingly endless public hearing testimony over the proposed nuclear power plant for Midland Dow to be built by Consumers Power Co.
>
> That has been added to the record this week with revelation that Midland County's Board of Commissioners has voted to spend $20,000 to promote construction of the plant. . . . This development was hardly necessary. . . . One need not have agonized his way through every day of the hearings to grasp the idea that the Mapleton people and the Saginaw Valley Intervenors have come prepared for a fight and prepared to see it through. Only reading of reports of the hearing would leave little doubt, moreover, that the power interests have been hard put and on the defense before the probing questions of Myron M. Cherry. . . .
>
> . . . .
>
> In Midland and in Washington, Cherry has argued in a way that has pricked the conscience of Congressmen, powers within the AEC structure and its licensing bodies.
>
> He has offered some solid evidence that not all has been or is yet being done in the best public interest. . . .
>
> . . . .
>
> It will be up to a judge to decide, of course, if it is proper for a public body to commit public funds to promote a private enterprise. There may be some way around this--but it is very difficult to see by what authority the Midland board indulged itself to this extent.
>
> . . . .
>
> It would seem that its duty was clearly to stay out of an already complex issue and leave the promotion to those who stand most to benefit from construction.[44]

The Bay City Times noted that:

> As a public body, as a body representing the public and committed to promoting its welfare, a county board ought to be looking at a nuclear power plant proposal with raised eyebrows

---

[44] Saginaw News, August 26, 1971.

> and a healthy, hearty, determined skepticism. If ever a "show me" attitude is appropriate, that's the time.[45]

The Atomic Safety and Licensing Appeals Board, chaired by Algie A. Wells entered the debate over production of documents in September, 1971 with a memorandum which held that "disclosure of the presently withheld documents would, in our opinion, be contrary to the public interest." "There was no formal showing by the intervenor of relevancy or need." The ASLAB held that disclosure of internal working papers with "information which may be of little or no significance" could produce "adverse effects on the regulatory process."[46]

By September 1971, the principal pending matters were the effectiveness of the emergency core cooling system (which was subsequently made the subject of generic rule making hearings) and environmental questions. The debate among the parties turned to questions of procedure in handling the environmental issues. Objecting to an ASLB requirement that intervenors must, at the outset, state their position on environmental issues, Mr. Roisman pointed out that

> This procedure totally ignores the fact that Applicant has the burden of proof in this case and therefore until Applicant states its position in detail and the factual bases for that position have been explored no statement of contrary issues can be formulated.
>
> . . . .
>
> Equally important in the development of issues by the Intervenors is the Staff analysis. The Staff is an advocate of the public and as such has the burden to fully explore these environmental issues and to take positions on them. . . . Intervenors are entitled to the benefit of the Staff analysis prior to taking a formal position on all issues and prior to completion of all discovery.[47]

---

45 Bay City Times, August 24, 1971.

46 Memorandum, September 3, 1971.

47 Statement of Issues by the Environmental Defense Fund, Inc., September 30, 1971, p. 2.

Mr. Cherry concurred:

> We believe it is neither fair nor permissible for this Board to impose the initial obligation as to environmental matters upon the intervenors. . . . The status of the environmental issues in this case is just as if the matter had been noticed for hearing without the availability of a final or draft detailed environmental statement. Insofar as environmental issues are concerned, intervenors may not be required to state their contentions or finally move for discovery in the absence of such a report.[48]

The Applicant's environmental report was filed in Fall of 1971. EDF stated some objections to the report and the Mapleton Intervenors began to file testimony and affadavits. This led to the following comment by Milton Wessel, attorney for Dow:

> We protest the continued pouring of miscellaneous materials into the hopper by the Mapleton Intervenors, in violation of every procedural and substantive rule in the book.
>
> . . . .
>
> One gets the impression that the Mapleton Intervenors don't really care whether the Board or the parties read this material, but are simply stacking the record with a mass of affidavits, letters, books, magazine articles and the like, which they can use if, as and when the matter is ever reviewed by another body.
>
> It is an old trial lawyer's tactic to load a hearing record with the kitchen sink by way of offer of proof or otherwise, from which an appellate record can be created which will at least obfuscate, perhaps confuse, inevitably delay and - sometimes - result in improper reversal.[49]

On January 7, 1972, the Staff made available their draft detailed environmental impact statement for comment. The final impact statement was made available in March[50] and evidentiary hearings were scheduled for May 17, 1972. Prior

---

[48] Letter from Myron Cherry to Arthur Murphy, September 30, 1971.

[49] Letter from Milton Wessel to Arthur Murphy, January 6, 1972.

[50] Final Environmental Statement Related to Construction of Midland Plant Units 1 and 2, U.S. AEC, Division of Radiological and Environmental Protection, March 1972.

to those hearings, the Environmental Defense Fund and attorney Roisman withdrew from the case. Mr. Roisman's petition for withdrawal indicated that the decision to withdraw was a policy decision of EDF based upon a reevaluation of EDF's legal commitments and was based upon the fact that

> . . . two other intervenors--Saginaw and Mapleton--are ably represented and concerned in part with the same issues as EDF. The significance of the issues raised by EDF persists and its withdrawal is neither intended to be a waiver of those issues in the proceeding with respect to other parties nor an indication that the issues should not be decided favorably to the position taken by EDF.[51]

By the Spring of 1972, the ECCS hearings were underway in Bethesda, Maryland and this presented scheduling problems for Myron Cherry, the attorney for both the Saginaw Intervenors and the Consolidated National Intervenors at the ECCS hearing. He noted that the Saginaw Intervenors were

> . . . represented only by myself, and without my full and complete participation my clients will be denied adequate representation.[52]

He explained the importance of the ECCS hearings and the necessity of his presence there. He asked that the date for the hearing at Midland be changed to accommodate his participation at the ECCS hearings. The attorney for Dow Chemical Company objected strenuously to this request:

> Over and over the Saginaw Intervenors have been advised by those supporting the application and even by the Hearing Board itself, that other committments will not be permitted to interfere unreasonably with the orderly progress of this litigation. Trial counsel can only be split so many ways, and a party who chooses to retain a trial lawyer of outstanding reputation whose time is accordingly necessarily limited, cannot thereby be permitted to interfere with the rights of

---

51 Petition for Withdrawal of the Petition to Intervene, April 26, 1972.

52 Letter from Myron Cherry to Arthur Murphy, April 15, 1972.

> the public, and of the other parties, to a prompt determination of the issues.[53]

(Intervenors stated privately the irony they felt characterized this situation in that Chairman Murphy's full-time employment is as a Professor of Law at Columbia University and he scheduled the hearings, they contended, to fit with his teaching obligations.)

CPC agreed with Dow's position and so too did the Board. Thus, only one intervenor group, the Mapleton Intervenors, was represented at the environmental hearings, while Mr. Cherry was occupied as attorney for the Consolidated National Intervenors in the ECCS hearings being held at Bethesda, Maryland by the AEC.

In their draft Proposed Findings of Fact and Conclusions of Law, prepared after the conclusion of the hearings, the Saginaw Intervenors note:

> We are more convinced now than we were two years ago that the real difficulty with analyzing a nuclear reactor lies in the overwhelming commitment that is made before no one [sic] really has an opportunity to make an analysis.
> . . . .
> While we are not prepared to state that the bias we urge is personal or peculiar to the Intervenors, we are prepared to say, as we have said before, that Board members have a long and successful relationship with the development of nuclear power; an inherent (almost genetic) feeling that a loss of coolant accident will never happen; and that any safety or environmental problem raised during the course of licensing hearings can be resolved at some point before it is too late.
> . . . .
> Did Dr. Hall come to this hearing with an open mind about alternatives to nuclear power? We think not. See, for example, Dr. Hall's pronouncement that it was the Intervenors who had to convince <u>him</u> of the lack of safety rather than what the law requires that the Applicant and Regulatory Staff prove objectively the merit of their assertions. . . .

---

[53] Letter from Milton Wessel to Arthur Murphy, April 20, 1972.

Did Dr. Goodman come to this hearing with an open mind about the alternatives to nuclear power? We think not. . . .

Even Chairman Murphy was susceptible to what we regard as the "occupational hazard" in the nuclear industry.

. . . .

What has happened is that the Atomic Energy Commission, so disturbed with its obligation to hold hearings on decisions already made and incapable of reversal, and so fearful of emerging as a proponent against public hearings, has reacted irrationally at every turn; and its agent, including the Board members here, have unfortunately not taken issue with such irrationality.

. . . .

We trust the Board members will not take umbrage at the tone of this submission. The remarks made herein are evidence of the frustration that one group of Intervenors has experienced before one administrative agency. The fault lies with the Commission, the industry and their promotional perspective.

. . . .

We would ask the Board to deny the issuance of a construction permit on the grounds that the Applicant has not demonstrated that the public health and safety will be protected and that an insufficient and inadequate environmental analysis has been made.[54]

These findings were never filed with the ASLB because, according to Myron Cherry's argument to the Atomic Safety and Licensing Appeal Board, there was no "real" hearing held in the Midland case because the Saginaw Intervenors were in effect foreclosed from considering the issues during the proceedings. Therefore, the intervenors felt no obligation to file proposed findings of fact and conclusions of law.

On December 17, 1972 the ASLB issued an Initial Decision authorizing the Director of Regulations to issue a construction permit to CPC:

80. Upon consideration of the record of this proceeding and the findings of fact and conclusions, the Board has determined that:

(1) In accordance with the provisions of 10 CFR 50.35(a):

---

[54] Draft of Saginaw Valley et. al. Intervenors' Proposed Findings of Fact and Conclusions of Law (no date).

(a) The Applicant has described the proposed design of the facility, including, but not limited to, the principal architectural and engineering criteria for the design, and has identified the major features or components incorporated therein for the protection of the health and safety of the public;

(b) Such further information as may be required to complete the safety analysis and which can reasonably be left for later consideration, will be supplied in the final safety analysis report;

(c) Safety features or components which require further development have been described and the Board has been assured the Applicant will conduct research and development to satisfactorily resolve any associated safety questions before the latest date stated in the application for completion of construction of the proposed facility.

(2) The proposed facility can be constructed and operated at the proposed location without undue risk to the health and safety of the public;

(3) The Applicant is technically qualified to design and construct the proposed facility;

(4) The Applicant is financially qualified to design and construct the proposed facility;

(5) The issuance of a permit for the construction of the facility will not be inimical to the common defense and security or to the health and safety of the public.

81. Upon consideration of the record of this proceeding and the findings of fact and conclusions, the Board has determined that in accordance with the provisions of Section A.11 of Appendix D, 10 CFR 50:

(1) The requirements of Section 102(2)(C) and (D) of the National Environmental Policy Act and Appendix D of 10 CFR 50 have been complied within this proceeding;

(2) The appropriate action to be taken is to authorize issuance of the construction permit.[55]

---

[55] Initial Decision, December 17, 1972, pp. 61-64.

The ASLB incorporated into its decision

> certain conditions, which are set forth in . . . the Staff's Final Environmental Statement (FES), included a modification of the water intake design to minimize first loss, the relocation of certain transmission lines, the conduct of ecological studies and development of surveillance programs, etc.[56]

It noted that:

> Applicant has agreed to all of these recommendations except . . . [one] which would have required applicant to "prevent any discharge which would result in increasing phosphorous concentrations in the river above 0.05ppm." As a substitute for that, Applicant has agreed to treat phosphates so that total discharge including laundry waste and start up waste based on the actual average will not exceed 35 pounds per day exclusive of the pond reconcentration of existing phosphate levels in the river. The Staff agreed to that substitution and the Board sees no reason why the proposal should not be satisfactory. The Board would like to note for the record that its conclusions are based on the understanding the Applicant will carry out the recommendations to which it has agreed.[57]

Two trade publications commented on the decision to recommend a construction permit:

> The Midland plant was announced in mid-1968 and an application for construction was filed in January 1969. Hearings began in December 1970 and concluded June 15, 1972, in what has been one of the longest hearings on record.[58]
>
> * * * * * * *
>
> Midland Plant licensing delays have added over $430-million to the cost of the Consumers Power project. The Babcock & Wilcox pressurized water reactor plant (two 850 Mw units) received a construction permit two weeks ago, five years after the project was announced. During those years the projected cost has risen from $267-million to more than

---

56 *Ibid*., pp. 59-60.

57 *Ibid*.

58 *INFO*, Vol. 54, December 1972.

> $700-million, because of greatly increased bond interest rates, cost escalation, and other factors. In addition, Consumers will have to spend millions on purchased power and installation of interim generating capacity to fill the gap caused by the Midland delay. Midland was originally scheduled to start up in 1974; now the first unit is scheduled for 1979.[59]

On January 7, 1973 Cherry filed a motion (and supporting argument) "to recall and revoke the Initial Decision In the Matter of Consumers' Power Company (Midland Plant, Units 1 and 2)" and to declare the ASLB biased. He asked that a new Board be created by the Commission which would hold a "new hearing on all relevant issues."[60]

Cherry's motion was based on the requirement of the Rules of Practice (10 CFR 2.7.8) that a presiding officer has the duty to conduct "a fair and impartial hearing." Cherry cited Texaco, Inc. v F.T.C., 336 F 2nd 754, (D.C. Cir. 1964) which stated that an administrative hearing "must be attended, not only with every element of fairness but with the very appearance of complete fairness."[61]

The object of Cherry's motion was an article by Chairman Murphy entitled "The National Environmental Policy Act and the Licensing Process: Environmentalist Magna Carta of Agency Coup de Grace?"[62] Cherry pointed out that "the article discusses in some detail the controversy surrounding the proposed Midland units."[63]

----

[59] Nucleonics Week, December 28, 1972, p. 4.

[60] Motion and Supporting Argument to Require the Presiding Officer to Recall and Revoke the Initial Decision and to Declare the Atomic Safety and Licensing Board Biased, January 7, 1971, pp. 1-2.

[61] Ibid., p. 5.

[62] 72 Columbia Law Review, 963 (October 1972).

[63] Affidavit of Myron M. Cherry in Support of Motion to Recall and Revoke Initial Decision on the Grounds of Bias, January 7, 1973, p. 3.

Chairman Murphy, Cherry contended, viewed NEPA in his articles "particularly as interpreted by Calvert Cliffs, as interfering with the production of needed energy and development of nuclear power and particularly the proposed Midland units."[64] He charged Murphy with prejudicial bias and addended his contention with items from the manuscript which he claimed show the prejudice, as well, of Drs. Goodman and Hall.

Cherry's motion was denied by the ASLB on February 1, 1973. The Appeal Board upheld the denial in its Memorandum and Order (ALAB-101) of February 10, 1973. (The basis for the Appeal Board decision is discussed in Section IVC of this report.)

Once the matter of bias was dispensed with by the Appeal Board, Cherry filed a motion (January 15, 1973) of some 156 pages of Exceptions of Intervenors to the Initial Decision of the ASLB of December 14. In it he raised issues of the fairness of the Commission's administrative proceedings and its obligation relative to the public's rights to participate meaningfully in the AEC licensing process. He also raised issues of radiological safety and certain environmental matters.

The Appeal Board (Alan S. Rosenthal, Chairman; John H. Buck, Member; William C. Parler, Member) met on March 14, 1973 to hear arguments on Cherry's motion. On March 26, the Appeal Board issued an interim judgment on the exceptions concerning quality assurance and quality control as it felt that

> prompt decision and action on our part are necessary. . . . We have concluded that the initial decision reflects an erroneous view of the ambit of a board's responsibilities

---

64 Ibid., p. 6.

> in the QA area and, as a result, inadequately deals with issues which have been raised in that area. . . . Since we find the record sufficient for us to resolve those issues and to impose conditions which will satisfy applicable requirements, we hereby do so, and are modifying the initial decision to reflect both the additional findings and the supplementary conditions which we believe are necessary on this record.[65]

The Appeal Board was concerned with ensuring that the quality assurance program would be properly implemented by CPC and by Bechtel, the architect-engineer. The Appeal Board made clear that

> it is not enough for a licensing board to satisfy itself that, _if_ implemented, the program . . . will adequately protect the health and safety of the public. At least where, as here, there has been a legitimate question raised in the course of the proceeding, the board must go on to inquire into whether there is, in fact, a reasonable assurance that the applicant and its architect-engineer _will_ carry out the program in accordance with its terms.[66]

It ruled that a complete report on quality assurance actions should be filed before resumption of construction activities to ensure that "work already performed and the materials now on the site are in satisfactory condition",[67] and that periodic reports on construction must be submitted to the AEC Regulatory Staff.[68] Consumers Power Company filed a report on May 25, 1973; but the Saginaw Intervenors indicated to the Appeal Board that they felt the report was incomplete,[69] and as of this writing no action had been taken on this contention.

In May, 1973, the Appeal Board ruled on the remainder of the exceptions to the Initial Decision. With the exception of the quality assurance and quality

---

65 _Memorandum and Order_ (ALAB-106), March 26, 1973, p. 2.

66 _Ibid_., p. 7.

67 _Ibid_., p. 16.

68 _Ibid_.

69 _Saginaw Intervenors Motion to Enforce ALAB-106_.

control conditions announced in its earlier decision the Appeal Board ruled that "we find no occasion to alter the result reached in the initial decision."[70] In its lengthy decision considering each of the intervenors exceptions in some detail, the Appeal Board had some interesting comments on the role of intervenors in AEC proceedings. In discussing the failure of the intervenors to file proposed findings of fact and conclusions of law which were requested by the ASLB the Appeal Board noted:

> Intervenors frequently enter Commission licensing proceedings, including this one, with broad and far-ranging contentions. They have been afforded, and have utilized, the full range of the Commission's discovery procedures. But, even after obtaining a plethora of information through discovery, they have presented limited direct evidence and have often confined their evidentiary cases to the conduct of cross-examination. In this proceeding, they affirmatively failed to file adequate proposed findings and conclusions, even though they were asked by the Licensing Board to do so. Now, after the Licensing Board has issued a decision with which they disagree in many respects, they file exceptions with this Board which challenge, *inter alia*, numerous factual findings of the Licensing Board.
>
> Participation in this manner, in our opinion, subverts the entire adjudicatory process. One of the justifications for public participation in that process is the additional input which intervenors may contribute to the identification and resolution of relevant issues, and the assistance they may provide an agency in fulfilling its statutory mandates. In short, the right of participation in an administrative proceeding carries with it the obligation of a party to assist in "making the system work" and to aid the agency in discharging the statutory obligations with which it is charged.[71]

Opponents have vowed to continue their fight despite financial hardship and the visible evidence that construction is proceeding.

----

[70] *Decision* (ALAB-123) May 18, 1973, p. 3.

[71] *Ibid.*, pp. 5-6.

VERMONT YANKEE NUCLEAR POWER STATION

The Vermont Yankee (VY) Nuclear Power Station is located on the Connecticut River at Vernon, Vermont. The station employs a boiling water reactor which has a generating capacity of 1593 thermal megawatts to provide 513 electrical megawatts of net power output. Most of the electricity produced by VY will be used for distribution through the New England grid to other utilities in the New England area.

The plant is located in a scenically beautiful area of farms, pasture land, and heavily wooded sections. The villages of Vernon, North Vernon, and South Vernon, immediately adjacent to the plant site, have a population of 1024 (1970 population); Hinsdale, New Hampshire, two miles directly across the Connecticut River from the plant, has a population of 3276; and Brattleboro, Vermont, five miles upstream along the river, has an urban area population of 21,294. Detailed figures on population distribution are not critical to this discussion but are contained in the references listed in this chapter. It is, however, of interest to note the location of the nearest house (1300 feet from the reactor building) and the nearest school (1500 feet from the reactor building) with an enrollment of 163. Within 25 miles of the site, approximately 80% of the land is undeveloped; most of the developed land is used for agriculture and dairy farming.

The Connecticut River (and the impounded section of the river at the site which is known as Vernon Pond) is sometimes used for canoeing and sport fishing but no commercial fishing is done on this section of the river. There is a paper processing plant upstream and a Reservoir Project 15 miles downstream which provides a domestic water supply for over two million people in Massachusetts.

The nuclear reactor is of the boiling water type; that is, it produces steam for direct use in the steam turbine. The nuclear steam-supply system and the turbine were supplied by the General Electric Company.

Heat in the circulating water system may be dissipated in three ways. In open cycle operation the heated water is pumped from the river through the condenser and discharged into the river. In closed cycle operation, cooling towers are used to dissipate condenser heat and the only water discharged to the river is the blowdown (i.e., the water removed from cooling towers in order to eliminate dissolved solids). The plant can also operate in a so-called "helper" cycle in which part of the water is circulated through the cooling towers in order to reduce its temperature before discharge to the river.

On November 30, 1966, Vermont Yankee Nuclear Power Corporation (VYNPC) filed an application with the AEC for "all licenses necessary for the construction and operation" of the nuclear power plant at Vernon.[72] The corporation is a holding company sponsored by ten New England utilities. The license application contemplated completion of construction and commencement of commercial operation on October 1, 1970.[73] (It might be noted that the application to the AEC was by no means the only step in the process of obtaining a license for the construction. The listing of government agency applications, permits and actions in the Vermont Yankee case covers five pages of the AEC's Environmental Impact Statement.)

---

72 Letter from Roger Coe (VY) to Director, Division of Reactor Licensing, AEC, November 30, 1966.

73 Vermont Yankee License Application, Part A, December 2, 1966, p. V-1.

In June of 1967, the Advisory Committee on Reactor Safeguards (ACRS) completed its review of the application. After mentioning certain safety-related items which it felt deserved special attention, the ACRS concluded its letter to AEC Chairman Seaborg with the statement:

> The Advisory Committee on Reactor Safeguards believes that the items mentioned above can be resolved during construction of the reactor and should be followed by the Regulatory Staff. On the basis of the foregoing comments, and in view of the favorable characteristics of the site, the Committee believes that the proposed reactor can be constructed at the Vernon site with reasonable assurance that it can be operated without undue risk to the health and safety of the public.[74]

The AEC requires hearings in construction permit cases and permits interested parties to intervene in these hearings. Public hearings were held on August 1 and 2, 1967 and September 6 and 7, 1967. Opposition to the issuance of a permit was limited at this stage. John Burgess, Town Attorney of Brattleboro, in a statement on behalf of the Town of Brattleboro to the hearing Board, expressed some concern with the plant concluding that: ". . . the Board of Selectmen wish to record themselves as wholeheartedly in favor of progress, but not at the expense of the safety of any of our people."[75]

On August 18, Mary Panzera sent a telegram to the AEC on behalf of the Connecticut River Anti-Pollution League which read in part: "Protest granting exemption for Vermont Yankee to build at Vernon because of inherent unsuitability of site. . . ."[76] This was followed, on August 28, by a "Petition for Leave to Intervene" which described the Anti-Pollution League as:

----

[74] Letter from N. J. Palladino (Chairman, ACRS) to Seaborg, June 15, 1967.

[75] Statement of the Honorable John S. Burgess in Behalf of the Town of Brattleboro, p. 3 (no date).

[76] Telegram from Mary Panzera to AEC, August 18, 1967.

> . . . an unincorporated voluntary association having approximately fifty members, who are concerned with the ecology of the Connecticut River Valley, the protection and preservation of its natural resources and the protection of its air and water from man-made pollution. . . .[77]

The AEC, in its notice of hearing on June 30, 1967 in the Federal Register, had stated that petitions for leave to intervene were to have been received by the Commission by July 17. Through its attorneys, the applicant, VYNPC argued that the petition of the Anti-Pollution League be denied because of the late filing.

On September 6, 1967, the Atomic Safety and Licensing Board, chaired by Valentine B. Deale,

> . . . dismissed the petition on the grounds that in its judgment the petitioner did not show good cause for failure to file the petition on time.[78]

Aside from several isolated letters from individuals and groups (such as the Vermont Federation of Women's Clubs), there was no organized citizen opposition to the issuance of the construction permit once the Anti-Pollution League petition was denied.

On December 8, 1967, the ASLB ordered that

> . . . subject to review by the Commission . . . Vermont Yankee Nuclear Power Corporation is authorized to construct the facility in accordance with the application and with the evidence and representations entered in the record; and the Director of Regulations is ordered to issue a provisional construction permit. . . . [G]ood cause not having been shown to the contrary, this Initial Decision shall be immediately effective.[79]

A provisional construction permit was issued by the AEC on December 12, 1967

----

[77]Petition for Leave to Intervene, August 28, 1967.

[78]Order Confirming Denial of Petition for Leave to Intervene by the Anti-Pollution League-Connecticut River Valley Chapter, September 18, 1967.

[79]Initial Decision, December 8, 1967.

and, after hearings on the financial qualification of the applicant, a construction permit was issued on August 12, 1970.

The absence of organized citizen opposition to the plant during the construction permit stage can be explained, perhaps, by the fact that, in 1967, citizen concern about the environment (and about the effects of technology in general) had not yet achieved the level reached two or three years later. Thus, few people were moved to express concern for environmental problems and even those who did possess the interest were less well informed than they would have been had the case arisen two years later. (Witness, for example, the fact that the Midland case, described earlier in this report, has been characterized by vigorous citizen participation, both in opposition and support of the plant, at the construction permit stage.) It can probably be argued that the best time to have an effect on the design, construction, and operation of a technological project is before large amounts of money and resources have been expended and the design and construction completed. Many of the problems involved in the Vermont Yankee case in 1970-72 arise from the fact that the plant had already been constructed, making evaluation of alternatives and alterations in design and operating plans difficult and expensive, if not impossible.

The State of New Hampshire, an early intervenor, went to the Federal Courts to obtain review of the order of the AEC which denied New Hampshire's exceptions to the ASLB decision to recommend the construction permit for VYNPC. In its petition to the United States Court of Appeals for the First Circuit, New Hampshire noted that it had attempted to introduce evidence concerning the effects on public health and safety of thermal pollution from the proposed facility. It argued that the AEC had refused to consider evidence and denied that it had jurisdiction to review this aspect of the case.[80]

The petition held that the AEC had erred in its interpretation of its responsibility. The State of New Hampshire case led to a ruling that the AEC had no jurisdiction in non-radiological matters. The ruling by the Court of Appeals seems to have been somewhat reluctant:

> We conclude that the licensing board and the Commission properly refused to consider the proffered evidence of thermal effects. We do so with regret that the Congress has not yet established procedures requiring timely and comprehensive consideration of non-radiological pollution effects in the planning of installations to be privately owned and operated.[81]

(The enactment of NEPA in 1969 and the subsequent Calvert Cliffs decision (1971) considerably broadened the AEC's environmental jurisdiction. This is discussed elsewhere in the report.)

Public hearings are not required by the AEC after the construction permit is granted. But, in late 1970, the AEC was the recipient of numerous letters from individuals and groups requesting public hearings on the issuance of an operating license for the Vermont Yankee plant.

At the same time the AEC, in Appendix D of Part 50 of its regulations, responded to the National Environmental Policy Act by holding that any hearings for which notices were filed before March 4, 1971 would not be required to include consideration of non-radiological issues. A formal petition requesting that notice of the hearing not be filed until after March 4, 1971 was filed on

---

[80] Petition for a Writ of Certiorari to the United States Court of Appeals for the First Circuit by the State of New Hampshire, April Term 1969 (no date).

[81] New Hampshire v. AEC, 706 F 2d 170 (1969) cert. den 395 U.S. 962 (1969).

February 5, 1971 by Attorney Anthony Roisman on behalf of Dr. and Mrs. Irving Lyon of Bennington, Vermont "and on behalf of all other citizens of New England who are concerned with possible adverse affects [sic] from the operation of the Vermont Yankee nuclear power plant."[82] Roisman argued:

> As of February 5, 1971, no notice for the Vernon Plant had been filed. It would be a gross abuse of the AEC's responsibility to the public if it determined to file the notice within the next thirty days and thereby eliminated the full exploration of environmental considerations with respect to a plant about which there has been a massive public and Congressional concern.[83]

Nonetheless, a notice of hearing was filed by the AEC on February 27, 1971.

The six issues prescribed by the Commission for determination were:

> 1. Whether construction of the facility has been substantially completed in conformity with the construction permit and the application as amended, the provisions of the Act, and the rules and regulations of the Commission;
>
> 2. Whether the facility will operate in conformity with the application as amended, the provisions of the Act, and the rules and regulations of the Commission;
>
> 3. Whether there is reasonable assurance (i) that the activities authorized by the operating license can be conducted without endangering the health and safety of the public, and (ii) that such activities will be conducted in compliance with the regulations of the Commission;
>
> 4. Whether the applicant is technically and financially qualified to engage in the activities authorized by the operating license in accordance with the regulations of the Commission;
>
> 5. Whether the applicable provisions of 10 CFR Part 140, "Financial Protection Requirements and Indemnity Agreements", of the Commission's regulations have been satisfied; and

---

82 Petition, February 5, 1971, p. 1.

83 Ibid., p. 3.

> 6. Whether the issuance of the license will be inimical to the common defense and security or to the health and safety of the public.[84]

On March 9, 1971, the ACRS letter on the application for operation of Vermont Yankee was sent to Glenn Seaborg and concluded, after mentioning several items of concern:

> The Advisory Committee on Reactor Safeguards believes that, if due regard is given to the items mentioned above, and subject to satisfactory completion of construction and pre-operational testing, there is reasonable assurance that the Vermont Yankee Nuclear Power Station can be operated at power levels up to 1593 MW(t) without undue risk to the health and safety of the public.[85]

Several parties filed petitions to intervene in the proceeding on the operating license. The State of New Hampshire filed, noting that it had:

> . . . a substantial interest in the outcome of this proceeding as the facility is adjacent to and discharges into the waters of the State of New Hampshire, to wit, the Connecticut River. The land underneath the Connecticut River to the natural water mark on the west bank, which is the Vermont side of the River, is owned by the State of New Hampshire. . . .[86]

The State continued:

> The applicant has evidenced an unwillingness to comply with the regulations and laws of the State of New Hampshire and has steadfastily refused to submit to the appropriate local and state agencies an application to discharge waste and other materials into the Connecticut River. The failure to be bound by the New Hampshire laws . . . is a showing that the plant's operation will have a significant adverse effect on the environment.[87]

---

84 _Notice of Hearing_, February 27, 1971.

85 Letter from Spencer H. Bush to Seaborg, March 9, 1971.

86 _Petition to Intervene by the State of New Hampshire_, March 22, 1971, p. 2.

87 _Ibid_., p. 3.

The State of Vermont also filed a petition to intervene noting that:

> Petitioner is informed and believes . . . that a granting of the license . . . will have an adverse effect upon and will be inimical to the health, welfare and safety of petitioner's freemen, citizens and inhabitants and will be likely to result in a significant adverse effect on the environment.
>
> Petitioner is informed and believes . . . that suitable, reasonable and proper conditions should be attached to and made a part of any license issued to applicant herein so that the fueling, testing, and operation of applicant's nuclear reactor . . . will not adversely affect . . . the health, welfare and safety of petitioner's freemen, citizens and inhabitants. . . .[88]

A third state, Massachusetts, also petitioned to intervene since Massachusetts

> . . . has a substantial interest in the outcome of this proceeding as the facility is adjacent to and discharges into the Connecticut River which after the point of discharge flows into and through the Commonwealth [of Massachusetts].[89]

Several citizen groups filed petitions to intervene. The Conservation Society of Southern Vermont (CSSV), the National Wildlife Federation, the Vermont Natural Resources Council and the Lake Champlain Committee jointly presented a petition. (CSSV characterized itself as "a private, non-profit Vermont Corporation" with 950 dues-paying members. Its purposes include "public education in conservation, the sciences and woodsmanship and the prevention of pollution."[90] The National Wildlife Federation is a non-profit national organization with two and one half million members. It claims to be

---

[88] Petition to Intervene by the State of Vermont, March 29, 1971, p. 2.

[89] Petition to Intervene by the Commonwealth of Massachusetts, (no date), p. 2.

[90] Petition to Intervene, (no date), pp. 1-2.

the "world's largest conservation organization."[91] The Vermont Natural Resources Council is the Federation's Vermont affiliate with about 1100 individual and organizational members. The Lake Champlain Committee is a non-profit membership association with about 1700 dues-paying members.) The petitioners submitted that

> . . . they are organizations who, by their activities and conduct . . . have exhibited a special comprehensive interest in the environmental, conservational, recreational and economic aspects of nuclear power development, and, further, that they are adequate, qualified and responsible representatives of that aspect of the public interest which falls within the ambits of conservation and environment.[92]

They called attention to various environmental factors and then contended that

> . . . the policies of the Atomic Energy Commission as presently formulated, encourage the construction and operation of nuclear power plants without adequate safeguards for the environment.[93]

They further contended that:

> The delegation of powers to the United States Atomic Energy Commission is unlawfully broad, thereby violating the Constitutional requirement of separation of powers. The Atomic Energy Commission has openly and notoriously engaged in not only the regulation but also the promotion and operation of nuclear power facilities.[94]

The Environmental League of the Connecticut River Valley, a private, non-profit corporation of approximately 200 members also filed as an interested party. The League's stated purposes included "public education in sound

---

91 *Ibid.*, p. 2.

92 *Ibid.*, p. 5.

93 *Ibid.*, p. 7.

94 *Ibid.*, pp. 8-9.

conservational practices and the environmental protection of the Connecticut River Valley."[95]

The Natural Resources Defense Council, Inc. (NRDC) was another petitioner. NRDC is a non-profit, tax exempt membership corporation whose purposes are:

> 1) To preserve, protect and defend natural resources, wildlife and environment against encroachment, misuse and destruction.
>
> 2) To conduct research and to collect, compile, and publish . . . information . . . concerning natural resources, wildlife and environment and to conduct public education programs. . . .
>
> 3) To take whatever legal steps may be appropriate and proper to carry out the foregoing purposes.[96]

NRDC "has appeared to defend environmental interests in numerous court proceedings and administrative rule making procedures and adjudications."[97] NRDC's primary interest in the Vermont Yankee case was in the consideration of the disposition of high level wastes,

> . . . including the environmental impact of transporting, reprocessing, and storage of such wastes during the period while they remain dangerously radioactive. . . .[98]

The most active, persistent and therefore, important citizen group in the Vermont Yankee case has been the New England Coalition on Nuclear Pollution (NECNP) and it too filed a petition to intervene. The group is an unincorporated membership association whose members include private individuals and organizations. NECNP

----

[95] Petition to Intervene, March 24, 1971, pp. 1-2.

[96] Petition for Leave to Intervene, April 5, 1971, p. 2.

[97] *Ibid.*, p. 3.

[98] *Ibid.*, p. 5.

> . . . seeks in this intervention to uncover all the underlying facts and assumptions which are relevant to the issues involved and to guarantee that the consideration of the issuance of an operating license and its eventual issuance or denial are conducted in conformity with the applicable law.[99]
>
> . . . .
>
> The Petitioner and its members will be adversely affected by the issuance or denial of an operating license for this plant if such full exploration does not occur because they may either in the former case be subjected to real but undisclosed damage to their health, safety or environment or they may, in the latter case, be denied the benefits of electric power generated by a safe, healthy, and environmentally desirable nuclear plant.[100]

NECNP was and is represented by Attorneys Anthony Z. Roisman, Gladys Kessler and Karin Sheldon.

During the course of the hearings, the various intervenors raised numerous issues. The New England Coalition identified and sought consideration of technological, environmental, and procedural issues such as:

1. Will the releases of radioactivity from the plant during normal operation or during any possible accident be kept as low as practicable? Will such releases be detrimental to the health and safety of the public or the environment?

2. Are the safety features of the plant adequately designed, constructed and tested to insure that they will prevent any harm to public health and safety in the event of an accident?

3. Should applicant be required to develop final plans for public evacuation in case of an emergency or at least state assurances that the necessary evacuation will occur?

---

99 Petition to Intervene, March 31, 1971, p. 2.

100 Ibid., p. 3.

4. Are there unresolved doubts as to the validity of the ECCS Interim Criteria promulgated by the AEC?

5. Should applicant be permitted to include a portion of the Connecticut River in the restricted or exclusion area (surrounding the plant where the radiation doses and concentrations of chemicals allowed are highest) without demonstrating that it has acquired absolute legal authority and the ability to expel anyone from this part of the river?

6. Will the thermal discharges from this plant during normal operation and including any blowdown from cooling towers be detrimental to the environment?

7. Will the creation of fog or ice or any hot or cold air during operation of the cooling towers be detrimental to the environment?

8. Will the use of this land for recreation purposes either temporarily or permanently due to the course of the plant be detrimental to the environment? The Coalition points out that the site location analysis should be sufficiently thorough to be certain that all relevant considerations and factors have been examined.

9. What will be the effects of reconcentration of chemicals and readioactivity in river sediments or biota? In particular, the AEC provides no studies or analysis to demonstrate the precise level of radioactivity which each plant, fish or other organism as well as river sediments will retain.

10. What is the effect on fish, birds and other animals in the area, or on users of the river and nearby residents of noise produced in operation of the cooling towers?

11. Is the plant absolutely necessary in terms of the power demands of the residents of New England. Has applicant examined viable alternatives to operation of the plant, i.e, attempting to reduce the need for electricity through conservation of the existing energy supply?

12. Are the provisions of 10 CFR Part 20 and Part 100 relating to maximum permissible radiological releases valid?

13. Is the draft environmental statement in compliance with the requirements of NEPA? The Coalition argues that the statement's many inadequacies require the AEC to prepare a new statement that is more responsive to the issues raised.

14. Is it permissible under NEPA to postpone a complete assessment of the environmental impact of the operation of this nuclear power plant until after the plant has begun operation?

NEDC was concerned primarily with a single question: what is the extent of the AEC's jurisdiction with regard to the transport, disposal, and storage of high level radioactive wastes?

The State of Vermont professed "benevolent neutrality"[101] on the question of whether or not the plant should be operated. Vermont, however, raised questions such as:

1. How will the plant affect the conservation of future development for appropriate recreational and commercial use of the Connecticut River?

2. How will the natural beauty of the state be desecrated by utility development, including insufficiently planned powerline corridors?

---

[101] Interview with John Calhoun, August 15, 1972.

Vermont and the other two states involved were concerned with whether the plant's waste and thermal discharges will violate intrastate and federal water quality standards.

The AEC's draft detailed environmental statement was made available in February of 1971 and comments on the statement were received for several months thereafter.

The Calvert Cliffs case was decided in July of 1971 and, in effect reversed the decision in New Hampshire v. AEC which was discussed above. The decision prompted a letter from Anthony Roisman (who was also Counsel for the Calvert Cliffs Coordinating Committee, Plaintiff in the case) to Samuel Jensch, Chairman of the ASLB which was to hear the VY operating license case. Roisman commented:

> With the resolution of the Calvert Cliffs case the Gordian knot which has been tied in this and other nuclear power plant licensing proceedings has been removed. The decision in that case makes explicit . . . the obligation of the Atomic Energy Commission with regard to environmental matters. matters.
> It is now clear that every aspect of the environmental impact as well as other costs and benefits of this plant must be explored. . . .[102]

Hearings were held on August 10 before a hearing board consisting of Samuel Jensch, David Hall, and Ira Zartman.[103] One interesting limited appearance was made by Richard Wilson, a physics professor at Harvard who commented that

> I make this limited appearance because I feel that the Final Safety Analysis Report and the AEC's environmental report are

---

[102] Letter from Anthony Roisman to Samuel Jensch, August 5, 1971.

[103] Mr. Jensch is Chief Administrative Law Judge for the U.S. Atomic Energy Commission; Dr. Hall is a physicist affiliated with the Los Alamos Scientific Laboratory; and Dr. Zartman is a physicist who was formerly chief of the Reactor Physics Branch, U.S. Atomic Energy Commission.

> too technical to be understood by a layman. I feel that the issues can and should be presented more simply. I have now examined the final safety analysis reports for three power stations, as well as four environmental reports and they all have the same unintelligible format. Some of the questions put forward by the intervenors have brought out intelligible answers; I will use these in my presentation.[104]

Dr. Wilson, essentially a proponent of the use of nuclear power, then proceeded to discuss, in layman's terms, matters of radiation effects, power plant safety, and other issues involved in analyzing the effects of a nuclear power plant on its surroundings in an attempt to clarify these issues for public edification.

In October 1971, VYNPC submitted a "Motion for Issuance of a License Authorizing Initial Fuel Loading and Low Power Start-Up Testing." The license requested would permit operation of the plant at not more than 1% of full power; the motion held that operation at such a low level "would not give rise to any significant adverse impact on the quality of the environment." The applicant contended in a somewhat puzzling paragraph that:

> because construction of the facility is now substantially complete, the limited operation of the facility at not more than 1% of full power for the purpose of testing would not preclude the subsequent adoption of alternatives in the facility design or operation of the type which could result from a review pursuant to the requirements of the National Environmental Policy Act of 1969. Inasmuch as adoption of any such alternative would require modification of an existing facility, operation of the facility for the limited purpose of testing would not, in and of itself, preclude such modifications.[105]

Early in November a controversy over the possible use of political pressure on the AEC by U.S. Senator George D. Aiken (R-Vt.) flared. Senator Aiken,

---

[104] Limited Appearance, August 10, 1971, p. 2.

[105] Applicant's Motion for Issuance of a License Authorizing Initial Fuel Loading and Low Power Start-Up Testing, October 22, 1971, p. 1.

a member of the Joint Committee on Atomic Energy, in a letter to AEC Chairman James Schlesinger, noted that:

> Governor Deane Davis of Vermont is requesting the AEC to make an early decision on the application of the Vermont Yankee nuclear plant to begin its production of power or at least a limited amount of power at an early date. Provided that the information which you have requested from the owners of the plant is furnished promptly and assures safe operation, I support Governor Davis' request.
>
> It is my understanding that the largest producers and distributors of power in Vermont . . . are extremely apprehensive that insufficient power will be available to meet the requirements of the State in the very near future. While I cannot absolutely vouch for their fears, I am of the opinion that they are justified.[106]

This letter and press reports about its substance led to an exchange of letters between Anthony Roisman and the AEC;

> Roisman: A recent article in the Rutland Herald (Rutland, Vermont) alleges that substantial political pressure is being brought to bear on . . . the Atomic Energy Commission to authorize the operation of the Vermont Yankee Nuclear Power Station at 1% power. . . .
>
> We believe that the current public proceeding is the proper forum for consideration of these matters and are opposed to any attempt, no matter how well-intentioned, to circumvent those procedures. In the face of heavy political pressure, the Commission must persist in its resolve to decide the question of licensing nuclear power plants in open, orderly and public proceedings.[107]
>
> Shapar (AEC): This will advise you that no approaches to the Commissioners urging favorable action on this matter have been made. A permissible contact (i.e., regarding the status of the proceeding and urging an expedited schedule for consideration) has been made with the Commission. . . .[108]

----

[106] Letter from Senator George Aiken to Schlesinger, November 1, 1971.

[107] Letter from Anthony Roisman to Schlesinger, November 8, 1971.

[108] Letter from Howard Shapar, Assistant General Counsel for Licensing and Regulations, to Roisman, February 4, 1972.

> Roisman: Thank you for your letter of February 4, 1972 with reference to unauthorized communications between Chairman Schlesinger and parties to the . . . proceeding. I had hoped for a reply to my November 18, 1971, letter at a much earlier date.
>
> Your disclosure of the facts reinforces my fear that the Chairman of the Commission is being subjected to improper pressures. Senator Aiken in his letter . . . clearly indicates that he is acting for and on behalf of the State of Vermont which as you know is a party to this proceeding. The letter from Senator Aiken contains numerous allegations regarding the need for electric power in Vermont. This is a critical issue in the proceeding and hotly contested by many of the parties.[109]

Senator Aiken, in a hearing of the Joint Committee on Atomic Energy referred to Mr. Roisman's letters as:

> . . . clearly an attempt to intimidate Members of Congress and also the Atomic Energy Commission, although the Atomic Energy Commission stood up to him and told him that it was not a violation of the law for me to communicate with Mr. Schlesinger on paper and officially.[110]

Hearings on the 1% license were held in November and December, 1971 and January and March, 1972. On March 14, 1972 the ASLB ordered:

> . . . in accordance with the Atomic Energy Act . . . and the Ruels of Practice of the Commission that the Director of Regulation is authorized to issue to Vermont Yankee Nuclear Power Corporation an operating license authorizing fuel loading . . . and low power testing. . . .[111]

The 1% operating license was issued in March 1972. In May, hearings on the full power operating license continued. In July, 1972 the final Environmental Statement on the plant was issued by the AEC.[112]

---

109 Letter from Anthony Roisman to Shapar, February 9, 1972.

110 In AEC Authorizing Legislation FY 1973, Part 2, pp. 1111-1112.

111 Initial Decision, March 14, 1972, p. 24.

112 Final Environmental Statement Related to Operation of Vermont Yankee Nuclear Power Station, U.S. Atomic Energy Commission, Directorate of Licensing, July 1972.

On June 6, 1972, the Atomic Safety and Licensing Appeal Board consisting of Sidney Kingsley, John Buck, and Lawrence Quarles issued an order concerning a motion of the Natural Resources Defense Council which would:

> . . . open for consideration in the proceeding the environmental impact of spent fuel transportation and reprocessing, the disposition and ultimate storage of high level wastes, and the disposal of the contaminated structures of the nuclear generating station at the end of its useful life when it is decommissioned.[113]

The ASLAB

> . . . concluded that the Licensing Board should consider the transportation of spent fuel to a reprocessing plant, and the transportation from the nuclear power plant of high level wastes as well as of low level wastes, but should not consider in this proceeding operations of a reprocessing plant or the disposal of wastes resulting from reprocessing.[114]

The ASLAB order used 22 pages to justify this ruling which was based primarily on the argument that reprocessing facilities and the handling of wastes are not being "neglected" by the Commission and that examining of reprocessing and disposal in each of numerous pending licensing cases would lead to an enormous administrative burden as well as to needless duplication of effort.[115]

In an attempt to short circuit the licensing of several already constructed but not yet operating plants in order to avoid emergency power shortages, Congress passed, on May 17, 1972 Public Law 92-307 which is now Section

----

[113] Memorandum and Order, June 6, 1972, p. 2.

[114] Ibid., p. 5.

[115] In November of 1972, the Commission announced a two-day hearing on the environmental effects of the Nuclear Fuel Cycle, including the mining, nulling, enrichment, fabrication, reprocessing, waste management and transportation of nuclear fuel materials. The "legislative type" hearings were held on February 1, 2, 1973 in Washington, D.C.

192 of the Atomic Energy Act, as amended, which provided for the issuance of temporary operating licenses under expedited procedures. The ASLB in the Vermont Yankee case took careful note of the fact that "Section 192 is not a carte blanche for licensing action; the Congress enumerated precise conditions that must be met. . . ."[116] For issuance of a temporary operating license it must be found that:

> (1) The provisions of section 185 have been met [i.e., that construction has been completed]
>
> (2) . . . adequate protection of the environment . . .
>
> (3) operation of the facility in accordance with the terms and conditions of the temporary operating license is essential toward assuring that the power generating capacity of a utility system or power pool is at, or is restored to, the levels required to assure the adequacy and reliability of the power supply, taking into consideration factors which include, but need not be limited to, alternative available sources of supply, historical reserve requirements for the systems involved to function reliably, the possible endangerment to the public health and safety in the event of power shortages, and data from appropriate Federal and State governmental bodies which have official responsibility to assure an adequate and reliable power supply.[117]

On July 13, 1972, VYNPC filed a motion for issuance of a temporary operating license for operation at full power. NECNP, NRDC and the State of Massachusetts opposed the motion, the AEC staff supported it, the State of New Hampshire filed a statement of no opposition under certain conditions, and the State of Vermont assumed a position of "benevolent neutrality." A hearing was held in Brattleboro, Vermont on August 15 to consider the motion. The applicant did not submit information on historical reserve requirements as seemingly required

---

[116] Initial Decision, August 24, 1972, p. 10.

[117] Ibid.

under Section 192. "The explanation", states the ASLB, "was that it would not be meaningful to present such a power generating capacity of such a large pool as New England Power Pool."[118] The Board expressed great concern (and even anger) at the deliberate omission of the information. Newspaper accounts of the hearing document this. Under the headline "Vt. Yankee Hopes Dimmed at Hearing" the Brattleboro Reformer of August 16, 1972 noted that:

> Hopes of getting a temporary, full power operating license for the $190 million Vermont Yankee nuclear power plant in Vernon were dimmed Tuesday by the presentation of an apparent lack of data on the power needs in New England.
>
> . . . .
>
> The opponents [of the Vermont Yankee plant] seemed to get some support from the AEC Licensing Board when Jensch chastised Vermont Yankee for presenting what he considered an inadequate amount of data in the power need issue.
>
> . . . .
>
> He [Jensch] said he believed in strict adherence to the law and emphasized several times during the proceeding . . . that Vermont Yankee had not followed the law strictly. In the final minutes of the hearing tempers flared with Jensch almost shouting at Dignan [attorney for VYNPC].[119]

The Rutland Herald, in an article published the same day with the headline "Temporary License May be Denied Atom Plant", said that:

> It appeared here Tuesday that the request of Vermont Yankee Nuclear Power Corp. for a temporary license to operate its Vernon plant may be rejected because of the failure of the corporation to file a document proving the need for additional power in New England. The chairman of the U.S. Atomic Safety and Licensing Board . . . expressed great concern time and again with the lack of historical data on the reserve requirements of electrical power in New England. . . . [F]inally near the end of the long day Jensch lost his patience and shouted at Dignan.
>
> . . . .
>
> . . . [A]lmost every attorney at the hearing, with the exception of the Vermont attorney general's office, argued before the board that Vermont Yankee had failed to show there is a serious shortage of reserve power in the New England area.

---

118 Ibid., p. 12.

Nonetheless, on August 24, 1972, the Board ruled that:

> If the Commission determines that Section 192 does not provide including data on historical generating capacity reserve requirements as one factor, among others, to measure what generating capacity is needed from the temporary license sought, then the Board finds and concludes as follows:
>
> A temporary operating license shall issue in the form proposed . . . for a duration through February 28, 1972. . . .[119]

The question was certified to the Appeal Board which ruled that VYNPC did not have to file data on historical power needs and on September 8, 1972 the AEC issued a temporary operating license for 20% power operation (100,000 kilowatts) through February 28, 1973.

Attorneys for Vermont Yankee, Thomas Dignan and John Ritsher, subsequently explained to us that the phrase ". . . factors which include, but need not be limited to . . ." has traditionally been interpreted by the courts as meaning "such as" and therefore, they were not surprised by the Appeal Board ruling.[120]

This contention set forth in applicants' oral testimony and supported by Dignan's letter to Chairman Jensch of August 10, 1972, and brief of September 11, 1972 was upheld by both the ASLB and subsequently by the Appeals Board, as noted above. AEC staff attorneys also supported applicant's contention and, as noted the Temporary Operating License was granted. The Appeal Board (on September 5, 1972) noted that "under this language, the single finding required is that operation of the facility under the Temporary Operating License is:

> . . . essential toward insuring that the power generating capacity of a utility system or power pool is at, or is

----

119 Initial Decision, August 24, 1972, pp. 20-21.

120 Interview with John Ritsher and Thomas G. Dignan, Jr., attorneys for VYPC, March 2, 1973, Boston, Massachusetts at Firm of Ropes and Gray.

> restored to, the levels required to assure the adequacy and reliability of the power supply. . . .
>
> The prescribed finding is that operation is essential toward insuring the specified levels of power generation.[121]

On September 15, 1972 AEC staff attorneys recommended that the ASLAB "appropriately modify the initial decision so as to authorize the issuance of a temporary operating license at 100% of rated power."[122] The Appeals Board agreed and on October 12, 1972 granted a full power (100%) license for Vermont Yankee.

Late in October, with the plant already operating under the Temporary Operating License, the ASLB held hearings on the environmental aspects of the operation of Vermont Yankee. The hearing lasted four days and was marked by the only direct case presented by NECNP. Two expert witnesses gave testimony on the effects of cooling tower operation and the effects of plant discharge on aquatic life in the Connecticut River. (In contrast, AEC introduced some 18 scientific witnesses to the environmental aspects of the hearing alone.) The states of Massachusetts, New Hampshire, and Vermont also testified about effects on the river ecosystem, the latter adding testimony on the setting of the mixing zone and cooling tower effects.

On November 9, a one-day session was held in Washington, D.C. on AEC's cost-benefit analysis. The burden of intervenor cross-examination was carried by the National Resources Defense Council.

---

[121] Memorandum, p. 3.

[122] AEC Regulatory Staff's Brief in Support of Exceptions of Applicant and in Opposition to Exceptions of Intervenors to Initial Decision on Motion for Temporary Operating License, September 15, 1972, p. 19.

Almost from the start of operation Vermont Yankee has been plagued with problems. On November 8, 1972 a fire destroyed an auxiliary transformer and the plant was shut down for two weeks. During the period of shut down it was reported that radioactive emissions from the "off-gas" release stack were higher than expected due to hydriding of the fuel cladding. The Rutland Herald reported:

> There now is a considerable history of fuel problems at the Vermont Yankee plant during its short life. Late in the summer it was revealed that some faulty nuclear fuel had been delivered to the atomic plant by General Electric Co.
>
> The bad fuel was shipped back to GE for inspection.
>
> It was replaced and sent back to Vermont and then some of it went back again to GE. More problems had been discovered in the fuel.[123]

Subsequently, other incidents occurred. The series of mishaps in Vermont Yankee's various stages of fuel loading and initial operations are presented in some detail in a motion filed by attorneys Roisman and Sheldon, entitled New England Coalition on Nuclear Pollution Proposed Findings of Fact and Conclusions of Law (December 19, 1972) as well as in Applicant's Proposed Findings of Fact and Conclusions of Law in the Form of a Proposed Initial Decision and Also Constituting Initial Brief of Applicant (November 28, 1972).

There seems, in both documents, to be considerable agreement as to events, but disagreements concerning the importance, magnitude and implications to be assigned these events.

---

123 Rutland Herald, November 25, 1972.

The applicant noted that

> During the course of this proceeding, a good deal of evidence was adduced concerning various problems with respect to the fuel in the initial core. . . . After the fuel had been originally shipped to the site, it was discovered that there were deviations in enrichments. As a result, all of the fuel was shipped back to the manufacturer in early 1971. . . .[124]

After the fuel was returned, it was discovered in January of 1972 that nine fuel rods were still "out of specification on enrichment"[125] and in March of 1972

> . . . an additional 16 rods had to be replaced because a final check of records had indicated the possibility that these rods were improperly located in the core.[126]

In November 1972 it was discovered that

> . . . although unexpected due to improved fabrication techniques, it appears that the Vermont Yankee fuel has a problem known as hydriding which is caused by excessive moisture in the fuel pellets. This was indicated by higher than expected gaseous releases during periods of operation above 50% of power.[127]

The applicant also noted that on November 9, 1972 "a fire occurred in the main unit auxiliary transformer requiring a plant shutdown."[128]

Roisman (for NECNP) added that

> The following unplanned events have occured and the plant has not performed as predicted:

----

124 Applicant's Proposed Findings of Fact and Conclusions of Law in the Form of a Proposed Initial Decision and Also Constituting Initial Brief of Applicant, November 28, 1972, p. 18.

125 Ibid.

126 Ibid., pp. 18-19.

127 Ibid., p. 19.

128 Ibid.

> 1) Human error resulted in a failure to monitor off-gas activity at the outlet of the air ejectors. . . .
>
> 2) Fuel rods twice returned to General Electric for examination and subjected to allegedly intensive quality assurance nonetheless are now suffering from hydriding abnormality which has increased off-gas activity at Vermont Yankee almost equal to the administrative limit.
>
> 3) Applicant exceeded Tech[nical] Spec[ification]s limits of 10% of 10 CFR Part 20 gas limits when airborne particulate activity rose . . . due to turbine leaks which were improperly exhausted during an on-site normal power loss.
>
> 4) Radwaste tanks contained activity levels more than 20 times those permitted by Tech. Specs.[129]

Roisman added several additional points concerning violations of the technical specifications, the unavailability of the auxiliary transformers, inappropriate emergency traffic control and evacuation plans, fuel subject to fuel densification, radioactive releases not maintained as low as practicable and potentially faulty electric type valve operations.

On February 27, 1973 the

> Atomic Safety and Licensing Board authorize[d] license for full power operation provided that the existing cooling towers are initially used for year round operations, rather than used as proposed for a few months of the year. Board also approves, after modification, Staff's cost-benefit analysis even though some costs and benefits were not expressed in monetary terms. One modification to Staff analysis authorized use of average retail rate for electricity in order to express in monetary terms the benefit of electricity. A second modification considered local taxes as a benefit.[130]

In granting the operating license to the Vermont Yankee Nuclear Power Corporation for full term (40 years), full power operating (1,593 MWt) of the

---

[129] <u>New England Coalition on Nuclear Pollution, Proposed Findings of Fact and Conclusions of Law with Supporting Brief</u>, December 19, 1972, pp. 3-4.

[130] <u>Initial Decision</u>, February 27, 1973, p. 1.

Vermont Yankee facility, the Board found that with respect to radiological health and safety and the common defense and security:

1. Construction of the facility has been substantially completed in conformity with the construction permit and the application, as amended, the provisions of the Act and the rules and regulations of the Commission.

2. The facility will operate in conformity with the application, as amended, the provisions of the Act, and the rules and regulations of the Commission.

3. There is reasonable assurance (i) that the activities authorized by the operating license can be conducted without endangering the health and safety of the public, and (ii) that such activities will be conducted in compliance with the regulations of the Commission.

4. The Applicant is technically and financially qualified to engage in the activities to be authorized by the operating license in accordance with the regulations of the Commission.

5. The applicable provisions of 10 CFR Part 140, "Financial Protection Requirements and Indemnity Agreements", of the Commission's regulations have been satisfied.

6. The issuance of the license will not be inimical to the common defense and security, or to the health and safety of the public.[131]

and with respect to environmental matters:

1. The requirements of Sections 102(2)(C) and (D) of the National Environmental Policy Act and Appendix D of 10 CFR 50 have been complied with in this proceeding.

2. The Final Environmental Statement, as modified in this Decision, should be affirmed.

3. Having independently considered and balanced the conflicting environmental factors contained in the record of the proceeding, the Board concludes that the appropriate action is to authorize issuance of the operating license

---

131 Ibid., pp. 51-52.

> requested subject to the conditions set forth in the ORDER below.[132]

The Board did attach certain conditions to the License. The following, exerpted from the Initial Decision, are particularly notable:

> A. Applicant shall operate the nuclear power facility with closed cycle operations at all times until data are developed and determinations made respecting significant adverse impacts on the environment by the advisory group for non-radiological environmental monitoring, which group is established by section 2.2.2 of the Technical Specifications dated September 9, 1972 as further modified by Staff change dated December 27, 1972.
>
> B. The total residual chlorine concentration will be limited to 0.1 ppm in the immediate vicinity of the plant discharge, and such limit on chlorine discharge shall continue until evaluation of test results, which may permit a change if approved by the Commission.
>
> C. The Applicant will define a comprehensive environmental (chemical, biological, and thermal) monitoring program for inclusion in the Technical Specifications for the Station, which is acceptable to the Commission for determining changes which may occur in land and water ecosystems as a result of plant operation.[133]

Additionally, the Board outlined a series of reports that Vermont Yankee would have to make to Massachusetts concerning discharges of radioactive materials, sampling of discharges, monitoring, chemical and radioactive analysis, quantities of radioactive releases, transportation of nuclear materials and emergency notification.

On March 19, 1973, Attorney Roisman filed, on behalf of NECNP, "Exceptions to the Initial Decision" on Vermont Yankee. His objections were based on the following issues, among others:

---

132 Ibid., p. 52.

133 Ibid., pp. 53-56.

It was error to exclude any presentation of evidence to demonstrate that the Criteria For Emergency Core Cooling System for Light Water Reactors (Interim Policy Statement) if met by Vermont Yankee would not provide sufficient protection for the public health and safety.

It was error to conclude that the environmental effects of the uranium mining, fuel manufacturing, fuel reprocessing and waste disposal caused by the operation of Vermont Yankee was not part of the environmental impact of Vermont Yankee and did not have to be included in the Vermont Yankee Final Environmental Statement or be the subject of the hearing.

It was error to exclude from the costs of the operation of Vermont Yankee the adverse consequences which will occur from the use of electricity generated by Vermont Yankee.

It was error to exclude from the alternatives to the proposed full-time, full-term operation of Vermont Yankee, the use of energy conservation methods which could reduce the need for Vermont Yankee.

It was error to refuse to strike the Final Environmental Statement where the Final Environmental Statement failed to disclose the basis for exclusion of the Class 9 accident consequences.

It was error to refuse to strike the Final Environmental Statement where the Final Environmental Statement failed to disclose the existence and content of reliable opinion contrary to the Staff conclusion that a Class 9 accident is incredible and that compliance with the ECCS Interim Criteria provides a sufficient basis to assume that the ECCS will operate.

It was error to conclude that the proposed operation of the plant after February 28, 1973 and until an improved system to control gaseous effluents is installed in the Summer or Fall, 1973 represents a reasonable effort to keep releases as far below Part 20 as practicable where there was no evidence that the alternative of halting operation until the modification was complete was analyzed and where the need for Vermont Yankee during the interim period was not established.

It was error to refuse to reopen the record to allow the receipt into evidence of letters which relate to important matters of plant safety where the existence of the contents of the letters did not come to the attention of the NECNP until after the record was closed but prior to the issuance of an Initial Decision.

> It was error to conclude that the plant was shut down to correct leaking fuel rods or that the shutdown, if it did occur, was as a result of an "abundance of caution" by the Applicant.
>
> It was error to refuse to include in the license a condition that simultaneous service of all Staff-Applicant correspondence be made on the local Public Document Room in Brattleboro and on the NECNP.[134]

An Atomic Safety and Licensing Appeal Board chaired by Michael C. Farrar heard oral arguments on the exceptions in May 1973. The Appeal Board raised "sua sponte . . . the question of whether the applicant had adopted a satisfactory quality assurance program."[135] The Appeal Board's order of May 23, 1972 concluded

> . . . that the Board erred in refusing to consider, on their merits, the quality assurance problem and the safety matters raised by NECNP's motion.[136]

The Appeal Board remanded the proceeding to the Licensing Board for a determination whether the record should be reopened and whether a stay or modification of the initial decision was justified.

The Appeal Board, however, decided "that we need not vacate the initial decision" because it found that

> . . . none of the safety-related matters sought to be raised by the intervenors involves any threat to public safety of such imminency that it would require us to take immediate action on an incomplete record.[137]

Following considerable discussion of procedural matters involved in the remand and after a conference of the parties on June 11, 1973, the ASLB ordered

---

134 New England Coalition on Nuclear Pollution Exceptions to Initial Decision, March 19, 1973, pp. 1-8.

135 ALAB-124, p. 3.

136 Ibid., p. 4.

137 Ibid., p. 27.

that "the evidentiary record in this proceeding is reopened to permit presentation by the parties of evidence respecting the five safety contentions asserted by the New England Coalition on Nuclear Pollution"[138] since "the Licensing Board concludes that all items reflect significant safety consideration."[139] Those contentions dealt with the areas of quality assurance, defective fuel rods, fuel densification, thin walled valves, and pipe ruptures. The Board also decided that the determination of the Appeal Board that the safety matters did not affect the short term operation of the plant was sufficient reason to justify continued operation of the plant.

Both the applicant and NECNP have indicated their intention to file exceptions to the ASLB's order.

The stated goals of the NECNP were to make the plant safe and to inform the public on nuclear matters and their implications. The success of the latter is exceedingly difficult for us to judge as no survey research was undertaken as part of this study, nor would it have been possible in any meaningful way as our study began at a time when the hearings were already underway with a considerable amount of news print already expended on the controversy.

The former, though, is less difficult to make a judgment about. It is clear by all counts that the NECNP had no influence on the operation or design of the plant. Since the application for construction permit in December of 1967 two major alterations have been adopted. The first came about due to pressure brought by the state of Vermont for the construction of cooling towers

---

[138] Order Establishing Procedural Matters and Convening Hearing to Receive Evidence at a Date Later to be Designated, June 13, 1973, p. 13.

[139] Ibid., p. 14.

(two mechanical draft cooling towers) in hearings before the Vermont Water Resources Board.

The second, an advanced off-gas system, is presently being installed by VYNPC as a response to state pressures, according to attorney Dignan to facilitate compliance with Vermont limits for radioactive releases. The possible use of an interim off-gas treatment system until the permanent system next summer becomes operable might be attributed to the actions of NECNP, but if this is so, it is the only tangible result of the intervention.

On June 12, 1973, the Vermont Yankee Nuclear Power Station was struck by lightning.[140] A series of minor mishaps has continued to plague operation of the plant. And so it goes.[141]

---

[140] Rutland Daily Herald, June 15, 1973.

[141] Kurt Vonnegut, Slaughterhouse-Five, Dell Publishing Co., New York, 1969.

RULE MAKING HEARINGS: THE EMERGENCY CORE COOLING SYSTEM (ECCS)

Since the early years of nuclear power, the safety of nuclear power plants has been a matter of concern and interest for both the industry and its critics. The famous 1957 AEC report "Theoretical Possibilities and Consequences of Major Accidents in Large Nuclear Power Plants" (known as WASH-740) concluded that a serious reactor accident could have catastrophic results: 3,400 deaths at distances up to 15 miles; 43,000 injuries at distances up to 45 miles; and $7 billion in property damages. Whether the increased size of today's nuclear power plants combined with somewhat more sophisticated safety systems would lead to greater or lesser effects has never been established since WASH-740 has never been publicly updated.

In a nuclear reactor, the fission of uranium nuclei in the fuel creates heat which in turn, creates steam used to generate electricity. The operation of a nuclear reactor results in the accumulation of large quantities of highly radioactive fission products. These fission products also generate heat; enough heat to melt the fuel and its supporting structures unless adequate cooling is provided. During normal operation of a reactor, water flows through the reactor core removing heat from the core; the heated water is ultimately used to generate electricity and in addition the water cools the core and serves to prevent the melting of the fuel. In the unlikely event that the flow of cooling water should be interrupted either through a rupture of a pipe or other malfunction, the fuel would heat up rapidly and perhaps melt were it not for emergency cooling systems which are designed to flood the core with water. Meltdown of the core could result in the break of the reactor's pressure vessel and the release of high levels of radioactivity into the atmosphere. Since 1970 a controversy has developed over the effectiveness of the emergency core cooling systems in providing prompt and adequate cooling in the event of a loss of coolant accident (LOCA).

The possible effects of a loss of coolant accident are not completely known or understood. In 1963, the AEC began the construction of a reactor called LOFT at the National Reactor Testing Station in Idaho. LOFT referred to the mission of the reactor which was to perform a loss-of-fluid test. LOFT was described as the "largest and most vital" test facility in the AEC's safety program and the "focal point which provides a fundamental sense of direction to water reactor safety investigations." But, as Robert Gillette has pointed out:

> Nine years later, the AEC is still struggling to finish building the LOFT facility, let alone run it. At last report construction was 80 percent completed, but work is now more than 6 years behind schedule. LOFT's crucial experiments have been expanded in scope but postponed until 1974 or 1975; by then, as many as 80 nuclear power plants that might have made use of the results in their design will already be running.[142]

The effects of a core meltdown are not the only aspects of nuclear safety that are less than perfectly understood. In 1965, an AEC task force under W. K. Ergen of Oak Ridge National Laboratory pointed out the need for experimental investigation of the effectiveness of emergency core cooling systems. Those tests have been underway for several years. In the fall of 1970 a test of ECCS effectiveness in a pressurized water reactor resulted in a failure of the ECCS to provide water to the core and a failure of computer models to predict the test results. An AEC task force under Dr. Stephen H. Hanauer was

----

[142] Science, September 9, 1972.

established to review ECCS criteria and in June of 1971, the AEC adopted a set of Interim Acceptance Criteria, a set of guidelines for the operation of nuclear plants which would be in effect until further research on ECCS could be completed. The AEC noted in its interim policy statement that:

> The criteria for the acceptance of emergency core cooling systems have been developed in the concept of the defense-in-depth concept, with the primary assurance of safety being accident prevention, achieved by correct design, construction, and operation, and by adequate quality assurance. The loss-of-coolant accidents postulated in these criteria thus presuppose a highly unlikely event as a starting point.
>
> These criteria are applicable to all light-water power reactors except as otherwise provided. Improvements are expected in analytical techniques, and experimental programs are expected to provide increased and improved knowledge about ECCS performance. On the basis of such improvements in technology, these criteria will be modified from time to time.
>
> The Commission believes that these criteria for emergency core cooling systems provide reasonable assurance that such systems will be effective in the unlikely event of a loss-of-coolant accident.[143]

The June, 1971 criteria ordered a stepped up maintenance and monitoring program, and changes in the ECCS of some reactors. In addition, a set of operating guidelines was imposed for the purpose of limiting the consequences of a loss-of-coolant accident.

The Policy Statement noted that:

> Ideally, one would have available analytical methods capable of detailed realistic prediction of all phenomena known or suspected to occur during a loss-of-coolant accident, supported in every aspect by definitive experiments directly applicable to the accident. In the absence of such perfection, adequate assurance of safety can be obtained from an

---

143 <u>Interim Policy Statement: Interim Acceptance Criteria for Emergency Core Cooling Systems for Light-Water Power Reactors</u>, June 25, 1971, pp. 6-7.

> appropriately conservative analysis based on available experimental information.[144]

The performance of the emergency core cooling system was "judged to be acceptable if the calculated course of the loss-of-coolant accident is limited as follows":

> 1. The calculated maximum fuel element cladding temperature does not exceed 2,300°F. . . .
>
> 2. The amount of fuel element cladding that reacts chemically with water or steam does not exceed 1% of the total amount of cladding in the reactor.
>
> 3. The clad temperature transient is terminated at a time when the core geometry is still amenable to cooling, and before the cladding is so embrittled as to fail during or after quenching.
>
> 4. The core temperature is reduced and decay heat is removed for an extended period of time, as required by the long-lived radioactivity remaining in the core.[145]

In summer of 1971, intervenors in individual nuclear power plant licensing cases began raising the question of ECCS effectiveness. In order to avoid the duplication likely to result from the consideration of the issue in a large number of individual licensing hearings, the AEC announced a hearing on the interim policy statement to begin in January 1972 to provide a single forum for the discussion of the ECCS issue. George Freeman, an attorney for a group of utilities, described the reasons for the hearing.

> In response to the growing outcry from the nuclear power industry, the utilities, and indeed the environmental movement, the Commission turned to rule makings as an alternative to this mounting duplication and delay. There was strong feeling, particularly among the utilities, that

---

144 *Ibid*., p. 5.

145 *Ibid*., p. 8.

> the existing chaos in AEC licensing required the removal of generic issues, such as the validity of the Commission's ECCS . . . regulations, from individual licensing cases into national forums. There each issue could be resolved in a single proceeding devoted solely to that issue, with all the necessary witnesses and all interested parties present.[146]

A hearing board consisting of Nathaniel Goodrich, Lawrence Quarles, and John Buck was established to conduct

> . . . a legislative type rule making hearing for the purpose of aiding the Commission in its determination as to whether or not the subject interim policy statement should be retained in its present form or adopted in some other form.[147]

Parties to the hearings included reactor vendors (Babcock and Wilcox, Combustion Engineering, General Electric, and Westinghouse), a group of utilities (known, at the hearings, as Consolidated Utilities), three states which have played a minor role in the hearings, the Lloyd Harbor Study Group (a citizens group which has also played only a small role) and a group of some 60 citizen groups which had been active in local licensing hearings and who joined with a small, Cambridge-based group of scientists known as the Union of Concerned Scientists to form the Consolidated National Intervenors (CNI). (A list of member organizations is appended).

In its request to participate in the hearings CNI noted that:

> All Requesting Participants oppose the Interim ECCS Criteria upon the grounds that such Criteria are not adequate in light of the current state of the art of nuclear safety and are illegal pursuant to applicable law. . . .[148]

----

[146] George Freeman, "The Emerging Role of Rule Making in the AEC Decisional Process", April 17, 1972, p. 3.

[147] Scheduling of Public Rule Making Hearing, November 26, 1971.

[148] Consolidated Request to Participate (no date), p. 7.

The CNI list of contentions contained in the request to participate was based upon the analysis of James MacKenzie, Ian Forbes, Daniel Ford, and Henry Kendall of the Union of Concerned Scientists. In August of 1971, this group had written to the AEC to comment on the interim criteria and noted:

> The Union of Concerned Scientists believes that the minor modifications in design criteria for emergency core cooling systems and changes in routine inspection promulgated by the Commission is insufficient and inadequate response to the serious safety problems that have become obvious in recent months concerning the emergency core cooling system's unreliability.
>
> . . . .
>
> In a report issued July 26, . . . the Union of Concerned Scientists presented its assessment of the Idaho semiscale tests. . . . That report argued that these emergency core cooling systems would very likely fail to terminate the course of a loss-of-coolant accident. Inasmuch as the consequences of this system's failure during a loss-of-coolant accident would be a complete meltdown of the core and possible release of an appreciable quantity of fission products to the environment, the Union of Concerned Scientists responded to the evidence of emergency core cooling system deficiencies by recommending a moratorium on nuclear power plant licensing until safeguards of assured performance are provided and a review of the hazards posed by presently operating reactors by a qualified committee independent of the Atomic Energy Commission.[149]

CNI raised a number of issues in the proceeding including the following:

1. The lack of convincing evidence that the emergency core cooling systems for any plants will control a LOCA.

2. The inability of computer codes and analytical techniques described in the Interim Policy Statement to predict accurately the course of a loss-of-coolant accident, including the deviations from such codes that resulted in experiments conducted in the semi-scale LOFT Blowdown System at the National Reactor Testing Station.

----

149 Letter from Union of Concerned Scientists to AEC, August 25, 1971.

3. The relationship of various fuel element cladding temperatures to the ability of emergency core cooling systems.

4. The consequences of allowing the maximum fuel element cladding temperature to reach temperatures between 1200° and 2300° during a LOCA.

5. The existence of major gaps in basic knowledge concerning the nature and consequence of events during a LOCA which is due to lack of systematic and comprehensive experimental research in the area.

6. The history of the ECCS test program replete with examples of conservative assumptions made by the AEC which were later proven to be optimistic.

7. The effects of blockage on various components of the flow into the reactor core during a loss-of-coolant accident.

CNI was also concerned with a large number of procedural questions relating, among other things, to rules of discovery and cross-examination.

The opinions of vendors ranged from an assertion that the Interim Acceptance Criteria were sufficiently conservative to the contention that they were overly conservative and unnecessarily restrictive. The Consolidated Utilities supported the interim criteria though they indicated that they might be excessively conservative in view of the safety record of the nuclear industry.

The State of Maine which also participated, though in a minor role, sought clarification of what restrictions on power output of presently designed nuclear power stations would be required to meet the criteria. The State's interest was primarily in the effect that the criteria would have on the Maine Yankee nuclear power plant.

The procedural scope of the ECCS hearings which were originally described as "legislative-type" hearings was expanded even before the hearings began. In its "Supplemental Notice" the AEC indicated that

> All persons admitted as participants will present their testimony under oath and will be afforded the opportunity for relevant questioning of the witnesses of other participants as may be required for full and true disclosure of the facts.[150]

Thus, the hearing was given the trappings of an "adjudicatory-type" proceeding; this was to be an important factor in the protracted and often contentious hearing. George Freeman has commented on the changing rules and procedures of the hearings which he feels have provided a

> . . . major source of delay in the ECCS Rule Making. Procedurally . . . the rule making has evolved from day to day, from week to week. The rule making is not governed by any of the regular AEC rules of procedure. . . . Rather it is shaped by a continuing series of Commission and hearing board orders.[151]

In another paper he comments that

> . . . it seems clear that the current AEC rule makings are experiments to determine the practicality of consolidating and resolving widespread challenges to the validity of its regulations in such national hearings.[152]

One other interesting provision was included in the Commission's Supplemental Notice:

> At the request of a participant, the presiding board may permit a qualified individual who has scientific or technical training or experience to participate on behalf of that participant in the examination and questioning of expert witnesses.[153]

---

150 Public Rule Making Hearing: Supplemental Notice, January 6, 1972, p. 2.

151 George Freeman, "The Emerging Role of Rule Making in the AEC Decisional Process", April 17, 1972, p. 6.

152 George Freeman, "The AEC's Recent Experiment in 'Evidentiary' Rule Making", September 21, 1972, p. 4.

153 Public Rule Making Hearing: Supplemental Notice, p. 4.

An early schedule predicted that the hearing record would be completed by the week of March 21, 1972.[154] In fact, the hearing was in session for more than 100 days and has resulted in a transcript of 22,000 pages. Though much of the transcript is filled with procedural wrangling among attorneys, some interesting features of the ECCS controversy have emerged. For instance:

> Along the way the hearing has revealed that safety researchers and advisers to the regulatory staff had raised a number of fundamental objections to the interim criteria months before they were issued, and that they continued to press their criticisms right up to the start of the hearings.[155]

During the spring several nuclear safety experts from the national laboratories and the AEC regulatory structure testified that the interim criteria were not as conservative as the Interim Policy Statement indicated. For instance, nuclear engineer G. Norman Lauben noted that "there were certain portions of the [AEC's] testimony that I would have to consider personally as not being sufficient."[156] On March 9, Philip Rittenhouse of Oak Ridge testified that he and some of his colleagues had reservations about the reliability of ECCS. Morris Rosen and Robert Colmar of the AEC Regulatory Staff also criticized the interim criteria; their testimony prompted Robert Gillette of _Science_ magazine to comment that ". . . on the strength of his story, he [Colmar] and Rosen would seem to rank in the major league of government whistle-blowers."[157]

The availability and production of documents has been a matter of contention throughout the hearings; George Freeman notes that "AEC documents have

----

[154] _Order_, February 4, 1972.

[155] _Science_, November 3, 1972, p. 483.

[156] _Science_, May 5, 1972.

[157] _Ibid_.

come forth as a hesitant series of tidal waves." He feels that "this continuing stream of documents . . . strained the capacity of participants in the rule making to almost the breaking point."[158]

One interesting controversy over the availability of certain AEC documents arose early in the hearings. CNI requested some AEC internal working papers; the AEC staff counsel objected to the disclosure of the papers on the grounds that they were exempt under the Freedom of Information Act. At first the ASLB concurred, holding that "the refusal to release such internal working papers in no way prejudices the ability of participants in this proceeding to question the staff on the basis of its technical conclusions."[159] The ASLB continued:

> We believe the exemption of such papers contained in the Freedom of Information Act and the Commission's regulations was designed to protect internal documents that are fundamental to the ability of a public agency to reach technical conclusions such as those offered in staff testimony. The public policy underlying this claim of privilege is well recognized and is intended to prevent any inhibitions upon the open and frank exchange of views and opinions among Government personnel who are responsible for the formulation of conclusions, frequently in areas where there may be differences of opinion.[160]

On February 4, 1972 the Atomic Energy Commission in effect overruled the ASLB. The Commission addressed itself to the matter

> . . . because of the importance of the question and its relationship to the compilation of the record in this matter. While, as explained below, we have no difficulty with the premises enunciated in the Board's order, we believe, in the circumstances of this proceeding, that a different course is

----

[158] George Freeman, "Emerging Role of Rule Making in the AEC Decisional Process", April 17, 1972, p. 13.

[159] Order, January 31, 1972, p. 2.

[160] Ibid., p. 3.

> warranted, and are accordingly directing that the staff make available, with certain designated exceptions, the documents which are in controversy.
>
> . . . .
>
> While a blanket characterization is not possible, the documents contain, in the main, a mixture of data and opinion--in many instances inextricably interwoven; and the bulk of their contents are, as the staff contends and the Board held, properly exempt from automatic production. . . . At the same time, they may provide a basis for fuller exposition of the staff testimony in the instant proceeding. It is these two factors which the Commission balances here.

In conclusion, the Commission urged

> . . . all participants to remain mindful of the fact that we are not engaged in this proceeding in an adversary contest akin to a criminal trial portrayed in the popular media but rather are involved in a rule making proceeding dealing with matters of technical complexity involving different shades of opinion. A healthy measure of self-restraint is incumbent on all participants in this setting.[161]

The self-restraint dicta apparently did not take hold. An examination of the record of the hearing demonstrates that protracted legal controversy, often acrimonious and shrill, punctuated and indeed dominated the ECCS hearing. Accusations of "playing to the press" were freely made; raised voices and unloosed tempers were not uncommon. Hours passed between the posing of a technical question and its answer; the interim was characterized by innumerable objections, motions, and Board rulings which has led at least one attorney to complain of the "hopeless morass."[162] Levels of frustration on all sides were high.

The ECCS hearings went largely unreported in the press. An occasional article in the New York Times or Time Magazine and an excellent series of articles on ECCS and nuclear safety in Science magazine (a journal not widely

---

[161] Order, February 4, 1972.

[162] Interview with Myron Cherry, May 22, 1972.

read outside a small community of scientists and science policy experts) comprise the extent of media coverage. In August, Friends of the Earth published ten issues of a newspaper called ECO (which had its genesis at the June, 1972 environmental meeting in Stockholm) dealing with the ECCS hearings in an attempt to fill the void. ECO described the hearings as follows:

> For eight months, more or less, a side room in Bethesda, Maryland reserved by the Atomic Energy Commission has been the scene of protracted legal wrangling, hyperacute concentration on minutiae, and some of the longest pauses since Joshua made the sun stand still. Unfortunately, however, the outside world works in real time. So do the reactor builders. Despite periodic news-releases lamenting the "slowdowns" and "hold ups" the hearings are inflicting on reactor construction, reactor-sites show little sign of cobweb accumulation. The same cannot be said for the Bethesda hearing.[163]

Quite early in the hearing, the AEC took note, with some displeasure, of the course of the hearings. In its order of February 23, 1972, the Commission commented that

> The type of rule making hearing then initiated was recognized to be somewhat experimental in nature. Essentially legislative in character, it was an effort to develop a format which could bring to the rule-making process an enlarged means for direct public participation so as to contribute to the information available to the Commission in making its ultimate determination on the matter at hand. It was also intended to contribute to public understanding of the complex technical considerations involved.
>
> . . . .
>
> The Commission is concerned with the slow pace at which the hearing is moving and with the relative lack of consideration of substantive issues in the questioning to date.[164]

The ASLB, too recognized the difficulty:

----

163 ECO, Number 4, August 14, 1972, p. 1.

164 Order, February 23, 1972.

> Since the purpose of a rule making proceeding is to identify the fullest measure of data relevant to the issues, the presentation of testimony and conduct of questioning requires every reasonable opportunity to be pursued effectively. In this framework, the increasing frequency and disruptive effect of objections to questions put by questioning counsel has led the Board to the conclusion that objections . . . should be made in a context that minimizes their impact on the opportunity for comment which participants exercise through questioning. If, in fact, the volume and nature of objections adversely affect the ability of questioning counsel to develop evidence in the manner counsel believe appropriate, they might conceivably impair the right of the participant asking questions to present his statutorily authorized comment.[165]
>
> . . . .
>
> The Board's concern with delay arising from unnecessary procedural wrangling is supported by its informal review of the record which indicates that substantial portions have been occupied by such matters.[166]

CNI was represented by several attorneys during the proceedings, but the dominant figure was Myron Cherry. To aid in the preparation of technical questions, Cherry enlisted the aid of Daniel Ford, a 1971 graduate of Harvard where he majored in economics. Ford had worked with the Union of Concerned Scientists on their analysis of ECCS and himself wrote large sections of the CNI/UCS testimony. Early in the hearing, Ford, despite his evident lack of scientific credentials, was admitted by the ASLB as a technical interrogator as long as other parties did not object. For months, Ford conducted cross-examination of witnesses, earning the respect of the other participants at the hearing and of ECCS experts within the AEC and at the national laboratories. But in July, 1972, George Freeman, attorney for the Consolidated Utilities, raised an objection to Ford's participation and noted:

----

165 Procedural Rule on Objections to Questioning, May 17, 1972, p. 19.

166 Ibid., p. 27.

> In order to avoid setting any precedent for other proceedings, we wish to indicate for the record that at no time have we either agreed that Mr. Ford was qualified . . . or consented to his questioning our witnesses as a technical interrogator. Our position reflects no bias against Mr. Ford, who has acted with impeccable courtesy in all his dealings with us throughout this proceeding. Nor do we doubt his valuable assistance to you [Cherry].[167]

Freeman was concerned that permitting Ford to question witnesses might set a precedent for permitting "lay interrogators" to engage in cross-examination of technical experts in future AEC proceedings; he commented that "experience teaches that exceptions made under the duress of the moment often prove in time to become the rule."[168]

When Ford reappeared as one of a CNI panel of witnesses (with Henry Kendall), the Board sustained objections to his participation on the grounds that he was not qualified as an expert, except on those portions of the testimony he actually authored. When the question of Ford's admissability as a witness was certified to the AEC, George Edgar, attorney for Babcock and Wilcox, objected to even the limited participation of Mr. Ford. Arguing that the Board had had ample opportunity to observe Ford's competence and had ruled that he did not qualify as an "expert", Freeman claimed that the clear result of such a ruling should have been Ford's removal from the witness panel rather than the partial curtailment of his participation as contemplated in the Board's ruling on the matter. Friends of the Earth, in its newspaper, ECO, wryly commented:

---

167 Letter from George Freeman to Cherry, July 7, 1972.

168 *Ibid.*

> Mr. Ford . . . [has] a new degree in economics rather than nuclear engineering--that makes him a non-expert despite his year of intensive study of nuclear safety. Another Harvard economist, ironically enough, is chairman of the AEC.[169]

The AEC upheld the Board and the hearings continued into August.

The hearings recessed in late summer to permit the participants to prepare redirect and rebuttal testimony. The AEC's testimony represented an apparent concession to some of the critics of the June 1971 criteria; it proposed certain "conservative" changes in the criteria. Subsequently, members of the AEC Staff testified that they felt it

> . . . extremely unlikely that a plant that meets the suggestions contained in the Staff's supplementary testimony would in an actual loss-of-coolant accident get to the melting temperature. . . . The Regulatory Staff does indeed intend to convey that additional conservatisms in ECCS criteria are believed to be appropriate.[170]

L. Manning Muntzing, AEC Director of Regulation stated, "We find the added conservatism to be dictated by safety" while acknowledging that some existing plants might have to reduce output in order to meet the new criteria.[171]

CNI's redirect and rebuttal testimony dealt

> . . . not only [with] narrow questions . . . but we have also chosen to submit further testimony with respect to such fundamental issues as the necessity for a critical independent review of the subject evaluation models of the various vendors and the extent to which the Regulatory Staff's "track record" detracts from its assurances of conservatism with regard to the predictions of the improved evaluation models.[172]

---

169 ECO, Number 1, August 7, 1972, p. 8.

170 ECCS Daily Digest, November 17, 1972.

171 Science, November 3, 1972, p. 482.

172 CNI Redirect and Rebuttal Testimony, p. 1.0.

The testimony pointed out what CNI saw as gaps in the interim criteria and the need for more experimental and analytical information. It went on to note that

> The Regulatory Staff relies on the assertion that the evaluation models accepted by them, while not necessarily conservative in all their components are, in aggregate, suitably conservative in the ultimate predictions that they make. The extent to which this position of the Regulatory Staff can be considered defensible and rational involves the extent to which the Regulatory Staff can systematically assemble quantitative information on the existence of various conservatisms and ranges of various uncertainties, and can provide, in a meaningful, statistical estimate, a confidence limit for the codes. Far from doing this, the Regulatory Staff has stated its inability to quantify the margins that are presently available, this is especially disturbing in view of the fact that very many examples are available of so-called conservatisms in the Regulatory Staff's approach to ECCS that have proven on subsequent analysis or when compared with measurements to be unjustified and, in fact, not conservative.[173]

In January, 1973 the AEC released its draft environmental impact statement concerning the proposed rule making action. That statement concluded that the effects of the action on the environment would be negligible. On February 10, the Environmental Protection Agency submitted its review of the impact statement. It concluded that the "draft statement is inadequate." Among the deficiencies noted by EPA were the following:

> 1. It [the statement] is based on Emergency Core Cooling System (ECCS) performance criteria no longer thought to be sufficiently conservative by the AEC technical staff,
>
> 2. Unamplified, it includes an inappropriately support statement concerning the risks of Class 9 accidents, therefore, depriving the public of a quantified perspective on the risks being taken relative to this category,

---

173 <u>Ibid</u>., p. 5.1.

3. It lacks sufficient quantification of criteria related to ECCS reliability, and

4. It does not include a meaningful cost-benefit analysis.[174]

A more final impact statement, more substantial than the initial draft statement, was issued on May 9, 1973, and hearings on environmental matters were held in the summer of 1973. At time of publication no final decision had been issued.

---

[174] Letter from Sheldon Meyers, Director, Office of Federal Activities, EPA to Lester Rogers, Director of Regulatory Standards, February 10, 1973.

CHAPTER IV: ANALYSIS AND INTERPRETATION

PROCEDURES

The Atomic Energy Commission asserts that its public hearings are designed to serve three purposes: to provide an orderly process for the licensing of nuclear power plants, to raise and review scientific and technological issues, and to provide a means for public participation. Experience does not indicate that a single proceeding can adequately serve all of these purposes, nor would evidence suggest that any of them is well served by the present system.

Prior to the late 1960's, the AEC hearings were viewed by the Commission, and the Joint Committee on Atomic Energy as a process to win public support and only incidently to resolve issues. Thus, the Joint Committee encouraged public hearings in the AEC licensing process because:

> . . . full, free, and frank discussion in public of the hazards involved in any particular reactor would seem to be the most certain way of assuring that the reactor will indeed be safe and that the public will be fully apprised of this fact.[1]

Although the hearings, in theory, opened the door to public participation in the licensing process, there was very little public interest of involvement at first. An AEC official has described the early hearing as "a charade."[2] No opposition to the issuance of a permit or license was heard and none was expected and the applicant and the AEC were usually the only participants. In 1968 and 1969 public interest groups began to take an active stand on nuclear power plants as the nuclear industry grew and it became evident that the plants had

----

[1] Harold P. Green, "Nuclear Safety and the Public Interest", Nuclear News, September 1972, p. 77.

[2] Interview with an AEC attorney.

significant negative environmental effects. These groups realized that active participation in the AEC hearings was the only way to make an input to the decisionmaking process, leading to the construction and operation of a nuclear power plant.

However, the characterization of the hearings as "charades" remains apt. Some feel that the emphasis on public acceptance rather than issue resolution still dominates the hearings:

> The AEC has regarded public participation in the licensing process as a means for procuring public acceptance of its safety determinations. Risks of nuclear power plants are never discussed in explicit terms; to the extent they surface at all, they are referred to only in passing. . . . A calculated effort is made to invite the public to assume that the fact that the plant meets AEC standards means that there are no risks. Residual risks remaining after these standards are met are totally ignored.[3]

Others are even more overtly skeptical; commenting on the ECCS hearings, a Friends of the Earth publication noted that:

> Ostensibly, at least, the hearings now in progress at Bethesda [ECCS] are intended to develop information on the basis of which the AEC might decide whether or not to revise its interim criteria (and if so, how) in order to better protect the public health and safety. But there is reason to believe that no matter what the hearings reveal, the AEC will resist (if it can) any substantial changes in the design and operating criteria that would inconvenience the reactor manufacturers. The long and expensive hearings could turn out to be a prelude to an AEC announcement that it is standing pat.[4]

It is quite clear that none of the participants are happy with the procedures; the applicants see the hearings as delaying procedures and showcases for the

----

[3] Harold P. Green, "Nuclear Safety and the Public Interest", Nuclear News, September 1972, p. 77.

[4] Stockholm Conference ECO: Atomic Reactor Safety Hearings, Vol. II, No. 3, August 11, 1972, p. 3.

display of problems associated with nuclear power rather than as mechanisms for understanding the scientific and technical issues involved. The citizen group opponents of power plants see the hearings as an elaborate attempt to give the appearance, but not the reality, of public participation; and the AEC itself, at least in the informal, off the record comments of its staff, sees the hearings as a process of questioning already-made decisions and of satisfying statutory requirements. A few comments will demonstrate the extent of dissatisfaction with the current process. John G. Quale, President of Wisconsin Electric Power Co. has noted that,

> As it now stands, when the Commission grants a petition for leave to intervene, the decision is tantamount to an injunction against the issuance of a license. . . .[5]

An ad hoc lawyers group, chaired by Robert Lowenstein, noted in a report to the Atomic Industrial Forum that the purposes of the hearings:

> . . . cannot be served most effectively by a classical adjudicatory or trial-type procedure. The whole apparatus of prehearing conferences, elaborate discovery, objections to evidence, rulings on evidence, cross-examination, initial decisions, exceptions and briefs--the typical paraphernalia of a trial--is wholly unsuited to reasonable consideration of the issues involved in a nuclear facility review.
>
> . . . [T]he trial-type procedure seems to be counterproductive. It can be endlessly manipulated to produce delay and prohibitive expense.[6]

Citizen intervenors and their attorneys also complain about the hearings.

---

[5] John G. Quale in Hearings before the Joint Committee on Atomic Energy on AEC Licensing Procedure and Related Legislation, June-July, 1971, p. 252. (Hereafter these hearings are cited as LH.)

[6] Atomic Industrial Forum, "A New Approach to Reactor Licensing in Joint Committee on Atomic Energy Hearings on Licensing Procedure", in LH, 1971, pp. 1447-1448.

David Comey of Businessmen for the Public Interest finds the hearings "inefficient and consumptive",[7] in effect, a waste of time.

Commenting on some of the procedural issues in the Midland case, Myron Cherry said:

> We cannot understand this series of events and are quite perplexed that the administrative process can so cavalierly operate to discourage (and even prohibit) public participation.[8]

In recent hearings before the Joint Committee on Atomic Energy, former AEC Chairman Schlesinger called the licensing hearings "happy hunting grounds" for lawyers and intervenors, and noted that the structure permits "delaying tactics, blackmail, and the like."[9] One thing is quite clear: the process as it is currently structured provides the means for neither the competent resolution of scientific and technical issues nor the adequate participation of the public in a manner which permits the consideration and resolution of problems perceived by the public and of concern to it.

Adversary proceedings are not an efficient way to resolve scientific and technical details. The hearings are legal proceedings, evidently designed by lawyers for lawyers and as such they are ill-designed to deal with the problems of arriving at scientific or technological truth.

Lelan F. Sillin, President of Northeast Utilities, has noted that:

> . . . the problems are frequently so complex the tribunal is not equipped to deal with them; decisions tend to

---

[7] Interview with David Comey, September 7, 1972.

[8] Letter from Myron Cherry to Arthur Murphy, May 24, 1972.

[9] James Schlesinger, testimony at Hearings before the Joint Committee on Atomic Energy on Nuclear Safety, January 23, 1973.

> become political and may interminably be delayed by lengthy hearings and appeals.[10]

The hearings are no more successful as a means of citizen participation. Though changes within the AEC in the last two years have tended toward the accommodation of citizen interventions, new policies seem to have attempted to extend procedural due process (altered discovery provisions, easy acceptance of citizens as participants) while not in any way changing the basic structure of the hearings to permit citizen group input to play an important role in the decisionmaking process. Citizen groups feel, and apparently rightly so, that they are being appeased but that they are not really participating.

The problems of using a single proceeding to serve two, perhaps disparate purposes were noted by Roger Cramton, then Chairman of the Administrative Conference of the United States, in 1971:

> Is the purpose of the hearing to reach an accurate determination of the particular issue of scientific fact or is the purpose of the hearing to satisfy the general public that they have had a chance to be heard and that their contentions have been fully aired and the like?
>
> If the objective is to have an accurate determination of a question of scientific fact, then presumably you want a staff investigation by the most expert scientists you can find who spend a great deal of time and use whatever method they think is appropriate to determine that question of scientific fact. Then they publish it, and disclose it and allow other people to attack it by comment and, perhaps, by oral argument procedures, but certainly not by a trial.
>
> But if what you want is public acceptance and confidence and the feeling that public intervenors have a chance to fully air

---

[10] Lelan F. Sillin, "Connecticut and the Energy Dilemna", speech prepared for RPI Colloquium on Balancing Environmental Concern with Economic Development, December 11, 1972, p. 1.

> their grievances and to test the positions that are being asserted against them, then maybe a trial serves a useful purpose.[11]

But the dual nature of the hearings is by no means recognized by all of those involved in nuclear power plant policymaking. Representative Craig Hosmer, a powerful member of the Congressional Joint Committee on Atomic Energy has said of the hearings: "If they do, in fact, contribute nothing in solving technical problems, we ought to get rid of them."[12]

The public hearing process has been cited by some as a major cause of delay in the licensing of nuclear power plants. John Quale of Wisconsin Electric Power Co. calls the hearing "a fertile ground for dilatory tactics."[13] Long Island Lighting Company (LILCO) attorney Edward J. Walsh believes that the hearings are a primary cause for delay:

> Although some people attribute the delays in AEC proceedings to be caused by the complex issues of licensing, I believe it is the practical problems in the hearing process which are the source of such delay.[14]

(It should be noted that LILCO has been involved in the hearings on the proposed Shoreham plant. Those hearings have been the most protracted hearings yet held on the issuance of a nuclear power plant construction permit.) Delay, in and of itself is not of particular concern to some of those involved in power plant licensing. Anthony Roisman wrote to Arthur Murphy during the Midland case that:

---

[11] LH, p. 353.

[12] Craig Hosmer in Hearings before the Joint Committee on Atomic Energy on AEC Licensing Procedure and Related Legislation, June-July 1971, p. 122.

[13] LH, p. 256.

[14] Edward J. Walsh, Jr., "Some Observations on Reactor Licensing Hearings", An Address at the Atomic Industrial Forum Conference, Tarpon Springs, Florida, April 19, 1972.

> We realize that unnecessary delay is to be avoided but were under the impression that any delay incident to a full consideration of the proper issues was in no way objectionable.[15]

The Atomic Safety and Licensing Appeal Board in the Vermont Yankee case argued similarly that when safety matters have not been resolved

> Any delay, . . . would be fairly attributable not to the intervenors but to the non-readiness of the facility for operation. Delay in the issuance of the license is entirely appropriate--indeed, mandated--in that circumstance.[16]

It is clear that the meaning of "necessary" or "unnecessary" delay is by no means agreed upon by all of those concerned.

But the hearings represent only one part of the licensing procedure and as such are only one possible cause for delay in the licensing of plants. While there is no question that some hearings have taken a considerable period of time to complete, it can be argued that delays due to the hearings represent only a small part of the time spent in processing and approving (or disapproving) permit or license applications. Arthur Murphy, chairman of the ASLB at Midland is concerned more with the potential for delay at the hearings than with the actuality:

> I emphasize that it is the <u>threat</u> of delay that is cause for alarm. Although the time spent in the licensing process is significant, there is, at present, little evidence that public participation in the process has been a major factor contributing to delays.[17]

James Ramey, AEC Commissioner, noted that there are several causes of delay other than the hearings:

----

[15]Letter from Anthony Roisman to Arthur Murphy, February 26, 1971.

[16]<u>Memorandum and Order</u> (ALAB-124), May 23, 1973, p. 26.

[17]Arthur Murphy, <u>The National Environmental Policy Act and the Licensing Process: Environmentalist Magna Carta or Agency Coup de Grace?</u>, 72 Columbia Law Review 963, 994, October 1972.

> . . . [T]here is no question that application review time has increased. . . . Many delays and increased costs have occurred as a result of other factors such as strikes during construction, problems in fabrication and construction, late delivery of components, and various technical problems cropping up during preoperational testing.[18]

He also attributed the "escalation of processing time" to "the proliferation of technical issues", the inadequacy of utility applications, and "the constant problem of not having enough staff to handle the application review load."[19] Edward Rogers, General Counsel for the Environmental Defense Fund, in discussing the time between application and granting of a permit or a license indicated that "it is my understanding that approximately four-fifths of such time is spent on pre-hearing review by the AEC Regulatory Staff and the Advisory Committee on Reactor Safety (ACRS)."[20] Staff members of the Administrative Conference of the United States completed a report on nuclear power plant licensing in April 1971[21] which concluded that while processing time from filing of an application to issuance of a construction permit more than doubled between 1966 and 1970, only some of the increase is attributable to more time spent at the hearing stage and that:

> . . . the major part of the increase stems from lengthening of the time required for staff and ACRS review. Generally, the total time spent on the hearing phase is only about one-fourth of the time spent prior to notice of hearing on pre-hearing analysis and review.[22]

---

18 James Ramey, Testimony to Joint Committee on Atomic Energy, June 22, 1971.

19 *Ibid*.

20 *LH*, p. 323.

21 Emory N. Ellis and James H. Johnston, *Licensing of Nuclear Power Plants by the Atomic Energy Commission*, Administrative Conference of the United States, April, 1971.

22 *Ibid*., p. 20.

The following table is taken from the Administrative Conference report and shows the amount of time spent in various phases of the licensing process.

Average Days of Processing Time at Various Stages of AEC Applications for Construction Permits for Nuclear Power Plants[23]

	1966[1]	1967[2]	1968[3]	1969[4]	1970[5]
1. Application to Notice of Hearing	214.0	251.7	344.1	411.8	460.8
2. Notice of Hearing to Hearing	37.0	36.4	42.9	47.4	41.1
3. Beginning to end of Hearing**	2.0	7.7	6.4	11.7	53.7*
4. End of Hearing to Initial Decision	12.3	34.5	40.3	36.3	52.0[x]
Total: Application to Initial Decision	265.3	330.3	433.7	507.1	607.6

* Includes time elapsed up until February 1, 1970, for cases in which hearing commenced in 1970 but for which Initial Decision has not been issued to date.

** Hearings are frequently interrupted -- this does not indicate the number of hearing days.

x Does not include data for 5 cases for which hearing has not been concluded.

[1] Figures for 1966 include 4 cases in which the hearing was started in that year.
[2] Figures for 1967 include 11 cases in which the hearing was started in that year.
[3] Figures for 1968 include 17 cases in which the hearing was started in that year.
[4] Figures for 1969 include 7 cases in which the hearing was started in that year.
[5] Figures for 1970 include 12 cases in which the hearing was started in that year.

---

[23] Emory N. Ellis and James H. Johnston, Licensing of Nuclear Power PLants by the Atomic Energy Commission, Administrative Conference of the United States, April, 1971, p. 21.

Although there are indications in more recent cases that hearing time continues to increase, the general point remains valid.

Now it is clear that the existence of the hearings does provide an opportunity for delay by any party which desires to use the hearings for that purpose. Only one group encountered in our study openly admits that its goal is delay. Wendell Marshall, a member and spokesman for the Mapleton Intervenors in the Midland case, has stated that his group wishes to stop the construction of the plant and will use all techniques available to it, including prolonged delay during the hearings, to achieve its goal. In our opinion, much of the questioning by the Mapleton Intervenors during the environmental stage of the hearings reflected the intention to delay the licensing procedure.

Despite the fact that delays in hearings are only a part of the reason for the increasing time spent in licensing review, it is clear that delays at the hearing phase are costly to all parties concerned. Estimates of the costs to a utility of allowing an already-built plant to lie idle after construction while operating license hearings are held range from \$1 million per month[24] to as high as \$3 million[25] or even \$5 million[26] per month. Delays at the construction permit phase are also costly. Delays at Midland (from all causes, including hearings) have added over \$430 million to the cost of the two-unit facility. In the five years between the announcement of the project and the issuance of

---

24 Bechtel Corporation, "Seminar on AEC Public Hearings", in LH, p. 414.

25 James Ramey, Speech to Atomic Industrial Forum, November 17, 1970, quoted in Ellis and Johnston, p. 35.

26 William Doub, Testimony before the Joint Committee on Atomic Energy, February 6, 1971, reported in Environment Reporter: Current Developments, Vol. 3, Number 41, February 9, 1973, p. 1222.

the construction permit, projected costs have risen from $267 million to over $700 million "because of greatly increased bond interest rates, costs escalation and other factors occurring during the time which the hearings have been held."[27] Utilities often blame citizen group intervenors for the delay and for the financial difficulties which accrue from the delay. James Campbell, President of Consumers Power Company notes that "Consumers Power Company has been one of the more noteworthy sufferers during the current rash of ractor licensing interventions."[28] Other parties also see the opposing intervenors as the perpetrators of delay. Milton Wessel, representing Dow Chemical Company, said in a letter to Midland ASLB Chairman Murphy that:

> . . . to this point its [the public's] right to an early determination--one way or the other--has been completely denied by the Saginaw intervenors' refusal to obey either the letter or the spirit of the Board's directions.
> . . . .
> How long must we--and especially the public--continue to accept and submit to the Saginaw intervenors' decision that there shall be no Midland plant?[29]

One cause of the increased time spent in hearings is the lengthy process by which intervenors identify and obtain documents and formulate and receive answers to interrogatories in the process of developing their case. Whether or not the time spent in waiting for applicants and the AEC to produce documents can be characterized as "delay" is open to question. Applicants see the requests for documents and the filings of lengthy lists of interrogatories as dilatory. Anthony Roisman, speaking for the Environmental Defense Fund in the Midland case, however, states:

---

27 Nucleonics Week, December 28, 1972, p. 4.

28 Letter from James Campbell to John Pastore, July 12, 1971, in LH, p. 438.

29 Letter from Milton Wessel to Arthur Murphy, February 25, 1971.

> . . . we cannot share the view of some parties to this proceeding which recklessly use the terms "unreasonable delay" as though the mere utterance of the terms provides an answer to any and all requests for discovery. The suggestion that public interest intervenors should become familiar with the Midlands [sic] nuclear plant in a few months when the Staff and applicant have had several years of familiarity with nuclear plants in general and this plant in particular is absurd. The development of that familiarity, the formulation of interrogatories and requests for documents, the analysis of responses to those requests and the development of a case for hearings is a lengthy process. It is axiomatic that so long as the AEC persists in allowing the public to intervene in proceedings only after the applicant and the Staff have completed their review and are in fact ready to issue a construction permit and begin construction that delay will occur. But it is unconscionable for responsible parties who know the complexities of these matters to claim that such a delay is unreasonable.[30]

But the utilities are not the only ones hurt by delays in the hearing process. Citizen groups, constantly under financial pressure, find it difficult to raise the funds to pay for legal fees and expert advice. According to Harold Green, attorney for intervenors in the Maine Yankee case, the nature of the intervenors' strategy was dictated, in large measure, by the paucity of funds.[31] And Edward Rogers of the Environmental Defense Fund notes:

> The public-interest intervenors have limited time and money. Protracted hearings--lengthy legal arguments over evidentiary points, for example--work against the interests of the intervenors as well as those of the applicant. Perhaps an intervenor, out of frustration, has on occasion resorted to delay as a tactic of desperation. It is certainly not a common practice and is not one which most public intervenors . . . would indulge in.[32]

Some intervenors' attorneys blame the delays on the AEC and the applicant.

---

30 Letter from Anthony Roisman to Arthur Murphy, May 17, 1971.

31 Interview with Harold P. Green.

32 LH, p. 323.

Myron Cherry has stated:

> It is my experience in contested hearings that the delays do not come from irresponsible intervenors. Rather, the delays come from the inability of an administrative agency to respond to proper and significant questions of safety and the reluctance of a utility or reactor vendor to disgorge information dealing with critical safety issues.[33]

Without conscious effort on the part of the intervenors, the applicants, and the AEC, there is a real danger of the hearings being reduced to a war of attrition, a test of staying power in which the party with the most adequate resources will inevitably prevail, independent of the validity of its case.

The holding of contested hearings at both the construction permit and operating license phases has been criticized by many (both proponents and opponents) as a cause of undue delay in the licensing process. It is claimed that holding two hearings leads to unnecessary duplication and that a contested hearing for an operating license forces the utility to allow a completed plant to stand idle while issues which are properly resolved at an earlier stage are discussed and argued. To date the duplication of hearings has been warranted by constant changes in statutory and legal requirements; the Calvert Cliffs decision, for instance, came after construction permit hearings in the Vermont Yankee case and thus dictated the holding of at least environmental hearings at the operating license phase. In fact, many of the plants currently in the operating license hearing phase had their construction permit hearings before the adoption of environmental legislation and subsequent executive orders and agency criteria to implement the legislation. Thus citizen participation at the construction phase was minimal, as at Vermont Yankee, and the operating

---

33 LH, p. 403.

license hearings have provided the only real opportunity for citizen opponents to participate in the AEC proceedings. It would clearly be more expeditious to have a single set of contested hearings, before the construction permit is issued, to resolve technical issues and to provide for citizen input to a decision on the ultimate operation of the plant. Consideration of an operating license application could then be limited to answering the question of whether construction of the plant met the requirements of the construction permit. It is equally clear, however, that such a situation cannot be considered a reasonable means of considering applications as long as the courts and the Atomic Energy Commission itself are continually altering the ground rules under which licensing shall take place. The changing procedural and substantive rules not only make a single hearing system difficult to envision, but lead, as well, to delays in the current hearing process. Disagreements among the parties about the proper scope of the environmental responsibility of the AEC and the subsequent decisions of the courts in New Hampshire v. AEC and Calvert Cliffs clearly delayed the expeditious consideration of environmental issues. In a letter to Midland ASLB Chairman Murphy, Myron Cherry noted, for instance, that:

> When my clients intervened in this case, we asserted that NEPA was being disregarded and violated. The AEC, steadfast in its unrealistic refusal to accommodate its licensing procedures to a modern age, took a contrary position; and Consumers Power contributed to that backward view by supporting the position of the AEC in the hearing. That position by the AEC and Consumers . . . resulted in delays in consideration of environmental issues on any serious basis from November 1970 to the fall of 1971.[34]

The Calvert Cliffs' decision partially clarified AEC responsibilities in dealing with environmental problems, but continuing disputes in the ECCS hearings

---

[34] Letter from Myron Cherry to Arthur Murphy, April 15, 1972.

and subsequent dramatic changes in the rule making hearing procedure on the environmental effects of the fuel cycle indicate that the era of changing ground rules is by no means over.

Despite the requirement for two hearings (at construction permit and operating license stages) some issues continue to be ignored in individual plant licensing cases: adequacy of emergency core cooling systems, environmental effects of fuel reprocessing and transportation, ultimate disposal of radioactive wastes, etc.

The ECCS, radiation emissions, and fuel cycle hearings represent an attempt to remove recurrent generic issues from individual licensing cases and to consolidate the consideration and resolution of these issues in a single forum. This makes good sense from the point of view of administrative and procedural expediency and can help the AEC to arrive at general conclusions concerning general issues common to all reactors. Generic proceedings can also serve to simplify individual licensing cases since the points of contention in the licensing cases would be reduced to those which are peculiar to the particular plant in question. There would still remain issues to be resolved in the consideration of each plant since design, construction techniques, quality assurance programs, and operating characteristics differ and since siting and locational differences may have a significant effect on need for and operation of a plant. One major difficulty must be overcome, however, before generic hearings can be considered to provide a reasonable alternative to a series of largely duplicative individual proceedings. As noted earlier, the hearings attempt to accomplish the goals of resolving technological issues and providing a means of public participation. And, while the potential for accomplishment of the first might be enhanced by increasing reliance on generic hearings, it is by no means clear that progress toward achieving the second would result from such an action. Citizen groups usually get involved because there is a local problem which excites or concerns them and aggregate for the purpose of dealing with the essentially local problem. It is much more difficult to generate public interest in generic hearings (located, most probably, in Bethesda or Washington) than in a set of hearings on a particular, local issue. Thus, if extreme care is not taken, an increased reliance on generic hearings could easily become a means for frustrating citizen participation by considering important issues in a relatively inaccessible forum, for all intents and purposes hidden from public view.

Until such time as issues common to all nuclear plants are either resolved in generic proceedings or considered in all individual plant licensing hearings, no single hearing system will be considered adequate by citizen groups.

Further, at present, there is no mechanism which provides a realistic check on the applicant's fulfillment of commitments incurred during the construction permit hearings. The hearings at the operating license stage theoretically, at least, provide such an opportunity. But hearing Board chairmen in both the Midland and Vermont Yankee cases have commented about the lack of adequate follow-up to ascertain the fulfillment of commitments.

One matter at issue in hearings at all stages is a determination of which party must bear the "burden of proof." Opponent intervenors have held that the burden of proof is on the applicant; the applicant must demonstrate that the plant in question is safe and necessary. Most intervenors feel that the AEC has not followed this interpretation of burden of proof. Thus Myron Cherry has written, with respect to the Midland case, that:

> As is well known, the Applicant in an Atomic Energy Commission licensing hearing has the burden of proof. . . . In this case the licensing Board placed the burden of proof upon the Intervenors.[35]

In another document Cherry asserts that

> It is important to remind the Board and the parties that the intervenors contesting the issuance of a license are in a real sense defendants in this proceeding and it is the Applicant, the Staff and the supporting intervenors who must carry the day on their burden of proof.[36]

In response Arthur Murphy commented that

> This conception of the intervenor as a defendant is, we believe, misleading. Intervenors are not defendants; insofar as they claim to be representing the public interest, they have voluntarily assumed a burden of representation of the public interest--not their own--and that burden will not always best be served by assuming the posture of defendants.[37]

His order insisted that in certain instances, lines of inquiry followed by the intervenors must "be based on technical evaluation of available information." He noted that:

---

[35] Myron Cherry, Exceptions of Intervenors Saginaw Valley Nuclear Study Group, et. al. to Initial Decision of the Atomic Safety and Licensing Board, January 15, 1973, p. 19. Cherry calls attention to the Administrative Procedure Act, Section 552 et. seq. For instance, Section 556 indicates:

> Except as otherwise provided by statute, the proponent of a rule or order has the burden of proof.

He also references the AEC Rules of Practice (10 CFR 2.732):

> Unless otherwise ordered by the presiding officer, the applicant or the proponent of an order has the burden of proof.

[36] Letter from Myron Cherry to Arthur Murphy, Goodman, Hall, December 15, 1970.

[37] Order, March 3, 1971, p. 5.

> We realize that such a course will put a burden on intervenors; however, particularly for those intervenors who claim to be representing the public interest, we do not feel that the burden is too high a price to pay for the assertion of those rights. At the same time, the Board confesses some uneasiness about those intervenors who are asserting essentially private interests. It would seem appropriate in some instances to distinguish between them and the "private attorneys general." However even as to them, a totally uninformed inquiry will not be tolerable.[38]

The "unease" expressed by Murphy is well taken. Virtually all parties involved in the nuclear power controversy view their interests as consistent with the "public interest." Anthony Roisman, when asked why he and the groups he represents claim to operate in the public interest, has publicly stated that the AEC and applicants probably feel that their private interests further the "public interest" and that intervenors feel the same. But in fact all of these groups represent only that part of the public interest which is made up of their own private interests.[39] Thus the New England Coalition on Nuclear Pollution represents the interests of its several hundred members; those interests also form part of the overall "public interest."

Chairman Murphy's argument above also seems to deny the existence of a middle ground between "totally uninformed inquiry" and a shift of the burden for detailed technical analysis of the plant and its operation to the intervenors from the applicant and perhaps AEC. It is quite probable that informed questioning about the problems and issues involved in power plant siting can occur without such a shift in the traditional burden of proof. Milton Wessel, attorney for Dow Chemical Company, noted that

---

[38] Ibid., pp. 7-8.

[39] Anthony Roisman, Speech to American Nuclear Society Meeting, November 16, 1972.

> . . . the intervenors in opposition to the plant . . . assume the task of challenging the proposal. Their intention to perform this function was the sin que non sic] of their intervention. And, once they have raised an issue which the Board considers worthy of consideration, it is the applicant . . . that must prove the issue to the satisfaction of the Board. Of course, the Board retains its power to raise other issues--not raised by the opposing intervenors--and as to these issues as well the applicant retains the burden of proof.[40]

It is evident that this interpretation of the statutes and rules regarding burden of proof is not universally accepted. David Hall, a member of the ASLB at Midland, stated during the hearing: "It is not up to us to convince you [intervenors] that the reactor is proper. You must convince us that it is improper."[41] George Trowbridge of the firm of Shaw, Pittman, Potts, Trowbridge and Madden, testified before the Joint Committee on Atomic Energy:

> Far more discipline is needed in the conduct of contested hearings than is now provided in AEC regulations. In particular, intervenors should be required, as a condition of intervention, to define their specific contentions with allegations of facts, supported by affadavits, which if true would warrant the denial or conditioning of a license.[42]

One of the major procedural controversies in all the cases we have examined revolves around the availability of documents and other information. Much of the delay attributed to the hearing process can be attributed to legal and procedural hassles over production of documents and responses to interrogatories. As indicated in the chapter on AEC licensing, Myron Cherry has estimated that 40% of licensing hearings are taken up in wrangling over the production of

---

[40] Dow Memorandum Concerning Procedural Guidelines, Issue to be Excluded, and Dow's Role and Contentions, November 25, 1970, p. 5.

[41] Midland transcript, p. 1048, lines 11-13.

[42] LH, p. 334.

documents which are, in the end, released. Requests for data and information from utilities and the AEC are invariably referred to as "prodigious" or "unrealistic" or "unnecessary." In response, intervenors claim that it is impossible to make an adequate case without access to all relevant information. The procedural hang-up involves the definition of "relevant." To many of those representing all of the parties in AEC hearings, such procedural wrangles seem unnecessary. An attorney representing nuclear plant proponents, who asked not to be identified by name, indicated that he felt that the withholding of any documents from the intervenors represented a silly and strategically unsound action on the part of utilities and the AEC. He subscribed to what could be described as a "needle-in-the-haystack" theory of document and information availability. There are so many documents and so much data involved in the analysis of a nuclear power plant that even if there were some information which was potentially damaging to the AEC's or utility's case, it would probably be overlooked or not even noticed by an intervenor who would be faced with a veritable mountain of documents and very limited scientific or technologically capable assistance. In fact, he felt, the AEC and the utilities make their cases much more difficult by attempting to withhold information or documents, since the very act of refusing to release certain information raises suspicion about the information and leads, almost invariably, to prolonged wrangling. Even without full disclosure of this sort, much raw data are available to the general public and intervenors in the AEC document room. But as predicted by the "needle-in-the-haystack" theory, useful data are obscured by the plethora of paper and even a careful and exhaustive (and exhausting) search through the mounds of documents cannot be expected with any certainty to bring to the surface that information which would be useful in structuring an intervenor's case. Serendipity helps, but one cannot rely on it. Indeed, without a healthy measure of long and arduous searching by someone who knows what he is looking for and con recognize what it is when he or she finds it, even serendipity will not come into play.

But even if true full disclosure became the rule, there is no guarantee that citizen intervenors would be capable of understanding, interpreting, and evaluating the information made available to them. Intervenor attorneys have freely admitted to us that they often obtain documents which they cannot use because they are unable to find competent advisors to read, digest, and interpret them. These attorneys see the problem as a largely procedural one. They insist upon the right to obtain the documents and feel that this right is independent of any use or non-use of the documents. There are exceptions, notably in the ECCS case, where the full time efforts of Daniel Ford and the involvement of other members of the Union of Concerned Scientists permitted the detailed examination and use of many AEC documents. However, the method of obtaining information in that proceeding points to additional problems. If the most reliable method for obtaining useful information is through the anonymous mailings of documents from public-minded AEC scientists, then surely one must question the extent to which any spirit of disclosure of information (as mandated by the Freedom of Information Act) is being adhered to by officials in any agency which controls the flow of documents and which feels for one reason or another that documents should be withheld from public scrutiny.

As in other aspects of AEC proceedings, changing rules concerning discovery make it difficult to analyze the details of the Commission's discovery policies. The recently revised rules of practice for the AEC (which took effect in the summer of 1972 after all of the cases examined here were in process) permit parties to an adjudicatory hearing to obtain information by means of written or oral deposition, by means of written interrogatories, through "production of documents or things or entry upon land for inspection and other purposes, and by requests for admission."[43] Unless otherwise limited by the presiding officer, parties are generally granted the right to discovery "regarding any matter, not privileged, which is relevant to the subject matter involved in the proceeding."[44] The real question of interest is how these rules will be implemented and we have no evidence upon which to judge this. One potential constraint enunciated in the new rules is the requirement that "discovery shall begin only after the pre-hearing conference provided for in §2.751a and shall relate only to those matters in controversy which have been identified by the Commission or the presiding officer."[45] The pre-hearing conference referred to here is the conference designed to "permit identification of the key issues in the proceeding" and to "take any steps necessary for further identification of the issues."[46] Thus, filing of petitions for intervention and identification of issues by the parties involved must take place before discovery, and, therefore, before documents have been examined. Depending upon the operational interpretation of "steps necessary for further

----

[43] 10 CFR §2.740.

[44] Ibid.

[45] Ibid.

[46] 10 CFR §2.751a

identification of the issues", this requirement may or may not work a major hardship on intervenors. Read in one sense it permits considerable latitude for identification and introduction of issues subsequent to discovery; taken in another sense, it limits the parties to consideration of issues identified before discovery begins.[47]

The inability of intervenors to obtain information early in a given proceeding has, at times, caused great difficulty. In asking for the lifting of a deadline for written evidence on environmental matters at Midland, Myron Cherry commented on April 15, 1972 that the deadline was

> . . . a particularly harsh burden in light of the fact that adequate preparation for the intervenors over the long months we have participated in this case has been thwarted by the failure to let us in the process early. In effect, since June 23, 1971, the date of the Calvert Cliffs decision, the Regulatory Staff and the applicant have had access to substantial amounts of information to prepare their environmental case. We asked during all those months to be permitted environmental discovery, but were denied the opportunity. The realization is that the collective efforts of the utility and the AEC have taken nine months to prepare their

---

[47] We have made some attempt to examine the discovery rules of other Federal agencies but find that without a significant effort a legitimate comparison among various agencies cannot be made. This is so because the mere language of the rules often has only a marginal relationship to actual practice. What is clearly important is how the rules are put into practice. The operational interpretation of rules of practice can only be analyzed through a careful observation of ongoing proceedings, an endeavor for which this study group had neither the time nor resources. Thus, although the AEC rules provide for several alternative methods of discovery as mentioned above, while the Federal Power Commission rules provide for depositions and subpoenas but make no allowance for interrogatories, no real conclusions can be drawn without a considerable effort to see whether or how the differences manifest themselves when the rules are applied. It is true that FPC depositions are permitted at *any* time before the hearing is closed, whereas in an AEC proceeding, discovery can begin only after the prehearing conference; but again, operational interpretation is critical. One could continue, in this vein, to trace difference in wording in FPC, SEC, FCC, FTC, ICC, etc., and AEC rules, but a more careful study would be required to draw definitive conclusions about the import of the differences.

> position on environmental issues and yet we are asked to duplicate that effort in less than a month without having the information available. We cannot see the fairness in such a procedure.[48]

There is no question that the fact that citizen groups often do not receive information until quite late in the licensing process puts them at a distinct disadvantage for the duration of their intervention.

ROLE OF LAWYERS

In the nuclear power cases, citizen group acquisition and use of scientific and technical information often reduces to the lawyer's acquisition and use of such information. The administrative hearings of the Atomic Safety and Licensing Boards provide a focus (an essentially legal focus) for citizen group intervention which establishes a pre-eminent role for the lawyer. At the hearings, there is no necessity for a citizen group to possess detailed technical knowledge of the plant and its effects (such as that possessed by the Citizens Committee to Save Cayuga Lake in the Bell Station case); the lawyer becomes the focal point of the intervention and it is his command of the scientific and technical issues and his scientific and technical support which become a critical factor. This is a relatively recent development; John Holdren and Philip Herrera trace the ascendance of the lawyer to the participation of Myron Cherry in the Palisades power plant case in 1970:

> Indeed, Cherry might well occupy a pivotal point in antinuclear disputes: the replacement of technicians, such as those behind the Cayuga dispute, with lawyers.[49]

---

[48] Letter from Myron Cherry to Arthur Murphy, April 15, 1972.

[49] John Holdren and Philip Herrara, Energy, Sierra Club, San Francisco, 1971, p. 232.

This is not to say that citizen groups themselves do not use scientific and technical information. These groups use the information to contribute to or improve public information; they attempt to illustrate to the public the dangers (or in cases where citizen groups support plans to construct a particular plant, the benefits) inherent in nuclear power plants and to build, as the hearing progresses, additional public support for their concerns and additional financial contributions to support their efforts. Virtually every group interviewed sees public education as a major function. (The Mapleton intervenors in the Midland case are a conspicuous exception.) Thus, Mary Sinclair of the Citizens' Committee for the Protection of the Environment of Michigan combines her science writing background with her interest in the Midland plant to disseminate information concerning the potential risks and dangers of nuclear power plants through the publication of pamphlets and papers as well as by encouraging media involvement. Nathaniel Smith, a trustee and recent past president of the New England Coalition on Nuclear Pollution, also emphasizes the public education function of the citizen group. "The facts should be made public", he argues "in order to provide a countervailing pressure to the public relations effort of the utility."[50] And Irving Lyon, who helped to form the Bennington Committee for Survival (a member of the New England Coalition) in early 1969 characterized the Committee as a "public information group", which brought speakers to the area in order to "build up a knowledgeable community."[51] Even citizen supporters of the Midland plant (Midland Nuclear Power Committee) see their role as

---

[50] Interview with Nathaniel Smith, May 30, 1972.

[51] Interview with Irving Lyon, June 1, 1972.

educational; they have sponsored participatory workshops and have even flown a group of citizens to Oak Ridge National Laboratory for an orientation program on nuclear energy.

But the primary interaction between proponents and opponents of nuclear power plants occurs at the ASLB hearings, and it is in this legal/administrative context that the citizen group must depend upon its lawyer to make its case. Roderick Cameron, Executive Director of the Environmental Defense Fund, notes that "the lawyer tends to regard himself in his traditional role as a 'hired gun.'"[52] The citizen group provides the "leg-work"; it copies documents, provides transportation for the lawyer, sometimes housing and often secretarial services, keeps a newspaper file, raises the money, holds and attends public meetings, and obtains documents,[53] but it is the lawyer who structures the case and carries out the intervention. Indeed, Harold Green, a lawyer for several nuclear power plant opponent groups, says that "the intervenor becomes the captive of the lawyer."[54] Myron Cherry feels that, with some exceptions, citizens and citizen groups make little effort to determine or understand the technical details of a nuclear power case. They oppose the projected plant and hire a lawyer to handle the case as he sees fit.[55]

The pivotal role of the lawyer is formally recognized by the State of New Hampshire in the State hearings on the proposed Seabrook power station.

---

[52] Roderick Cameron, "Exchange of Ideas Between Scientists and Lawyers: A Cost/Benefit Analysis", Mimeographed paper, December 29, 1972.

[53] Interview with Mary Sinclair, May 17, 1972.

[54] Interview with Harold Green, June 5, 1972.

[55] Interview with Myron Cherry, May 15, 1972.

There, Donald Stever of the Attorney General's Office has been designated as "counsel for the public" and members of the public with an interest in the proceedings have been requested to make this interest known to Mr. Stever. (It is of some interest that this State hearing was by far the best attended of the nuclear power hearings we attended. The large attendance was most likely due to the apparent interest of a significant part of the local public in requesting early opposition to the diversion of limited beach area for the construction of a nuclear plant.)

It is clear, then, that a study of citizen group use of scientific and technical information must consider the ways in which lawyers obtain and use scientific and technical information. In a few cases, such as the Cayuga Lake case, the citizen group itself played a major role in the generation and interpretation of technical information. But, in the cases studied in detail here, as well as at Maine Yankee and most other nuclear power plant cases, the scientific and technical detail, if any, is the province of the lawyer and his (or her) scientific advisors.

That this has a critical bearing on the level and quality of scientific input in the hearing process is beyond question. Thus Milton Wessel, attorney for Dow Chemical Company in the Midland case,is on record as stating, "we are not scientists, we are retained by clients to represent their interests. . . ."[56] The truth, he claimed, results from application of the adversary process. And at the same hearings, Robert Lowenstein, representing Consumers Power Company, expressed his annoyance at questioning of scientists and

----

[56] Midland Transcript, May 18, 1972, p. 5564.

engineers about the design of the cooling system for the Midland Plant by claiming that the chance of a lawyer or a biologist (the technical advisor to attorney Irving Like was Dr. Lawrence Holcomb, a Creighton University biologist) finding something wrong with the cooling system is "about one chance in a billion."[57]

The distinct roles of lawyer, scientist, and citizen in the hearing process deserve comment. For the most part, the citizen is reduced to an observer, a non-participant who must depend upon others for a clear articulation of his interests. The lawyer who is called upon to represent the citizens' interest must depend upon his own (perhaps insufficient) knowledge of the technology or upon the aid of scientific advisors. Although these advisors are, at times, permitted to cross-examine witnesses, their role is often severely restricted by the hearing Board. Thus, when Dr. Holcomb strayed somewhat from strictly scientific questions in his cross-examination, he was sternly rebuked by ASLB Chairman Arthur Murphy: "I don't propose to listen to a lay questioner arguing questions of law."[58]

Since the lawyer is a focal point for citizen involvement in power plant licensing cases, it becomes important to consider the ability of lawyers to grasp the often complex issues of science and technology involved. Clearly, the grasp of issues varies with the particular lawyer. However, some generalizations can be drawn from our observation of several cases. None of the lawyers involved has a real (complete) understanding of technical details of the cases, although some possess a very clear view of the critical issues. Because

---

[57] Midland Transcript, May 19, 1972, p. 5963.

[58] Midland Transcript, May 18, 1972, p. 5817.

of this, there is a tendency during the conduct of a hearing to permit incorrect or misleading answers to stand without following up the points raised by the questions. The questions themselves may well have been written by a technical advisor who may or may not be present at the time. Anthony Roisman describes a case in which a misleading response to a question was not recognized, as such, by him, or, it should be added, by his technical advisor (who was indeed present but who may or may not have been the author of the question). Only a following question from a member of the ASLB pointed out the obfuscatory response which would have otherwise gone unnoticed.[59] In some hearings few, if any, of those present are sufficiently technically knowledgeable to know the difference. There is, further, a tendency to ask essentially random, groping questions in an attempt to expose an area in which scientific knowledge of the witness is vague and incomplete or to be satisfied with having evoked an equivocal response from a technical witness. Roisman admits that in the Indian Point case he freely used this technique of questioning, a technique which he feels should not be permitted in ASLB hearings. He felt that because of his lack of expertise, he "couldn't really focus on critical issues", and had to guess at which issues were important by "feeling around" and observing witnesses' demeanor.[60] Some seeming, though unclear, psychological advantage seems to be perceived by attorneys upon the exposure of the witness' ignorance on a particular point, independent of whether or not the point has real relevance in the particular case or comes within the scope of the expertise of that particular

----

59 Interview with Anthony Roisman, July 12, 1972.

60 <u>Ibid</u>.

witness. Moreover, this technique is not limited to the arsenal of intervenor attorneys, but is frequently observed being used by most lawyers in these cases.

Whether or not lawyers possess a complete understanding of the scientific and technological issues involved, it is clear that some of them are confident that they do and are willing to proceed in hearings without scientific or technical advise immediately available. Harold P. Green, attorney for the intervenors in the Maine Yankee case, is confident that he can handle the issues involved in the licensing hearings on that plant based on his long experience with nuclear power. He used his own knowledge to "hammer out the details of the stipulation" among the parties in the Maine Yankee case.[61] Lawyers on the other side of the nuclear power controversy show similar confidence. At the ECCS hearing, George Edgar, attorney for Babcock and Wilcox, was accompanied on the first day of testimony by the Union of Concerned Scientists (members of the Consolidated National Intervenors) by technical experts from B&W, but after two days of protracted legal haggling, he sent the experts home since "they have more important things to do."[62] He felt that he could competently handle the technical questioning of the witnesses because of his own engineering background.

On the other hand, some lawyers feel considerably handicapped by a lack of scientific expertise. Karin Sheldon, attorney for the New England Coalition, framed her own questions during the portion of the hearing on Vermont Yankee

----

61 Interview with Harold P. Green, May 25, 1972.

62 Interview with George Edgar, August 4, 1972.

devoted to radiological and safety issues. She stated that she would have felt better with scientific help and that, in fact, she was "out of [her] depth" without such help.[63] An AEC technical expert noted that it was Ms. Sheldon's first case and that she was "out in left field."[64] Interestingly enough, it is our opinion that Ms. Sheldon was quite competent in her questioning of witnesses, particularly on radiological issues, although on matters of engineering details she was on shaky ground.

Myron Cherry, attorney for the Saginaw Intervenors at Midland and the Consolidated National Intervenors at the ECCS hearings, admits his lack of scientific expertise but makes a point of being accompanied at hearings by scientific experts and of maintaining telephone contact with others. At the ECCS hearings, his technical advice (coming, as it did, from the Union of Concerned Scientists and from anonymous notes written by AEC employees) was remarkably competent and the exceedingly long proceeding left Mr. Cherry with a good understanding of the general principles involved. In the words of Daniel Ford of the Union of Concerned Scientists, the intervenors at the ECCS hearings are "on top of the scientific issues."[65]

Anthony Roisman, another attorney with extensive experience in nuclear power plant licensing cases, feels that his performance in those cases was hampered by his lack of understanding of the scientific and technical details and issues. Perhaps Mr. Roisman is unduly harsh with himself; one of his

---

63 Interview with Karin Sheldon, May 20, 1972.

64 Interview with Warren Minnars, May 31, 1972.

65 Interview with Daniel Ford, May 22, 1972.

scientific advisors in the Vermont Yankee case, Marc Goldsmith, found Roisman "quick to grasp"[66] the scientific and technical issues, and at least one of his clients is impressed with his ability to "sift and choose"[67] among scientific inputs when experts disagree.

On the other hand, Irving Like's questioning in the Midland case (even with a scientific advisor present) can best be characterized as vague and unstructured. The clear inadequacy of the questioning raises important questions about Like's grasp of the scientific issues as well as his ability to determine the adequacy of scientific advice he receives. Since scientific advice to intervenors is quite often inadequate, and because scientific and technological issues are often at the heart of the matters in contention, it is incumbent upon the lawyer to find ways to evaluate the advice he is getting and to use only that which has a substantial basis in scientific fact. Like has made his feelings on the attorney's role in the licensing process known in a paper entitled "Multi-Media Confrontation--The Environmentalists' Strategy for a "No-Win' Agency Proceeding." There he states that "a new lawyer-client relationship must be developed which reflects the nature of the struggle as a part legal, part political campaign."[68] He goes on to note that:

> . . . winning the ultimate environmental objective requires maximum use of the media and arts of communication in dramatizing the confrontation between the citizen and his corporate and agency adversaries, and that all of the skills of

---

66 Interview with Marc Goldsmith, June 20, 1972.

67 Interview with Diana Sidebotham, Vice President of NECNP, June 1, 1972.

68 Irving Like, "Multi-Media Confrontation - The Environmentalists' Strategy for a 'No-Win' Agency Proceeding", Presented at ALI-ABA Course of Study on Environmental Law, January, 1971.

> counsel and his dedicated lay and scientific allies must be exerted to the task of educating the public to understanding the nature of the particular technology and its environmental effects, and moving the public to adopt the ecological ideal.[69]

Almost without exception, lawyers tend to ignore the caveat of Roderick Cameron of the Environmental Defense Fund who warns:

> . . . The lawyer must realize the diversity and limitations of science. A biologist is not an automotive engineer, nor a traffic projection engineer, nor a flood control expert, nor a meteorologist, nor a nuclear engineer, nor a systems analyst. Yet a lawyer is prone to think that once he has allied himself with a technically trained person his "expert problem" is done. He should learn that his scientific parties must develop liaisons with a broad spectrum of applied scientists whose expertise will bear directly on the problem. As an expert, a scientist has a narrow input into any broad problem.[70]

Some participants in ASLB hearings blame many of the intervenors problems on some combination of the incompetence and often arrogant and contentious manner of their lawyers. This, they feel, unnecessarily polarizes the hearings and exacerbates disagreements and sometimes leads to ad hominem rulings by the Board. Our observations bear this out in some cases although in fairness to lawyers for intervenors, we have noted the same characteristics displayed at various times by lawyers for the AEC, the utilities, and the reactor vendors. Consequently, criticisms of licensing proceedings often reduce to criticisms of the tactics or personalities of the individual lawyers involved. An AEC official, for instance, attributes the difference between the relatively calm and expeditious "as low as practicable" hearings and the drawn-out and

----

69 Ibid.

70 Roderick Cameron, "Exchange of Ideas Between Scientists and Lawyers: A Cost/Benefit Analysis", Mimeographed paper, December 29, 1972.

acrimonious ECCS hearing to the personalities of the lawyers for the intervenors. Roisman, he felt, was interested in getting facts on the record in the ALAP hearing, while Cherry, in ECCS, had "no interest in moving the hearing along." Cherry tended to "play to the press" and was constantly on "the verge of a nervous shriek."[71] Commissioner Ramey had a similar interpretation of the lawyers' personalities and attributed the "failure" of the ECCS hearings to the presence of Myron Cherry rather than to the procedures used at the hearing.[72] A member of the ASLB panel concurred, noting that Cherry's "rude and arrogant" manner was a direct cause of much of the hearings' acrimony. And, still another ASLB member laid many of the "intervenor's problems" in one case to Cherry's demeanor before the Board.[73]

The ASLB hearings are designed by lawyers evidently for the convenience of the lawyers. They become lively and exciting when dealing with legal and procedural matters; they become dull to the extreme of boredom when technical matters are discussed. (We have even seen lawyers fall asleep during the course of hearings, and spectators disappear when dull technical details are being discussed.) A former AEC official feels that intervenors tend to test the agency process rather than question on technical fact. They seek, he feels, to show that the agency and its rules are confused or inadequate, rather than to examine technical matters.[74] (Once again, it is our observation that this

----

[71] Interview with an AEC official.

[72] Interview with James Ramey, November 6, 1972.

[73] Interviews with ASLB members.

[74] Interview with Martin Hoffman, October 3, 1972.

preoccupation with procedure over substance is characteristic of all the lawyers involved rather than only the intervenor lawyers.) One intervenor attorney thinks that the Midland case is an example of this; he has characterized that case as "all strategy, no substance."[75] Another attorney for nuclear power plant opponents stated that "the adjudicatory hearing process for the licensing of nuclear power plants is not viable, intelligent, economical or productive."[76] Intervenors, he felt, have been unable, in the process, to raise the real and serious outstanding technological issues that underlie the nuclear controversy. He pointed out that the role of the attorneys (for both sides) required them to keep any real questions or answers which diminish their case out of the record; neither side has any commitment to the full exposure of truth. The result is that cases are stated in extreme and polarized terms; it can be assumed that the truth lies somewhere in between. "Winning", he claimed, means different things for each of the parties involved. For the applicant, "winning" may be the obfuscation of the real issues; for the opponents, "winning" means making applicants (or vendors) look bad while casting sufficient doubt on even generally accepted technological facts so that the impression is created that the technology is dangerous. While this is not ordinarily achieved by the intervernors, protracting the hearings increases the potential for the ASLB to commit legal errors which will provide the basis for appeal to the courts.

We concur with the opinions expressed by many of the parties above. Non-substantive, procedural issues dominate the hearings, to no discernible public

----

[75] Interview with Anthony Roisman, July 12, 1973.

[76] Interview with Harold P. Green, June 5, 1972.

advantage, to no ultimate proof or examination of scientific fact. Martin Hoffman, until recently AEC General Counsel, isn't sure that the solution to the problem would not be to take the hearings out of the administrative law process, since, in his view, it may well be administrative law that causes the problems.[77] Joint Committee on Atomic Energy member Craig Hosmer (R-Cal.) has also expressed frustration with the predominance of "lawyers and the other non-producing people" in the hearing process.[78] Friends of the Earth took this wry view of the dominance of lawyers in the hearings on emergency core cooling systems:

> . . . [M]uch discussion centered on the matter of so-called "interrogatories" submitted by one or another participant. "Interrogatories?" What ever became of "questions?" Linguistic anarchy of this kind typified most of the discourse, underlining the surreal inconsequence of the proceedings. Sometimes the dialogue came across like missing pages from Lewis Carroll. Indeed, by comparison, "Who Stole the Tarts." looks like a model of rational coherence.[79]

The current system of ASLB hearings presents the observer with the irony of a group of lawyers taking part in hearings being run, ostensibly, to resolve scientific and technical issues, at least in part. It is somewhat bizarre to watch lawyers seeking to establish an evidentiary record on scientific matters through their questioning on issues they evidently understand barely, if at all. The result is that undue attention is paid to points which have little or no relationship to the issues at hand. Hearings are marked by frequent

----

77 Interview with Martin Hoffman, October 3, 1972.

78 LH, p. 430.

79 Stockholm Conference ECO: Atomic Reactor Safety Hearings, Vol. II, No. 4, August 14, 1972, p. 6.

interventions on procedural matters and other areas in which attorneys are evidently more comfortable. Long periods of delay result from legal gymnastics and the procedural wrangling obfuscates the essential issues at hand. And even the "technical questions" asked by the lawyers are frequently so basic and simplistic as to be meaningless in the context of the hearings. Thus, in the ECCS hearing, Dr. Henry Kendall of the Union of Concerned Scientists was asked to define Boyle's Law, a question perhaps appropriate on a freshman college examination but hardly helpful in the evaluation of ECCS criteria.[80] Questions are often unrelated and irrelevant; it is difficult to see how many of the "technical questions" asked could lead the attorney to any substantial conclusions or could help him make his case.

The ECCS record contains two almost incredible exchanges between AEC lawyer James Murray and CNI witness Kendall concerning documents which both had before them:

> Q: Is it your testimony that this document contains 22 pages?
> A: The document appears to have twenty-two pages. . . .
> . . . .
> Q: Is it your testimony that this document has two pages?
> A: Yes.
> Q: That's all I have, Mr. Chairman.[81]

One would be hard pressed to find any redeeming value in such nonsense. An observer has noted, somewhat facetiously, that the only ones to benefit from the intervention and the hearings are the lawyers for both sides, and that the lawyers for the applicant and vendors evidently benefit more adequately than

---

80 ECCS Transcript, pp. 17140-17141.

81 Ibid., p. 21716.

do the attorneys for intervenors.[82] AEC Chairman Schlesinger, upon his departure from the Commission, characterized the hearings as providing a "happy hunting ground" for lawyers and characterized the representation of intervenor groups as "a lucrative substitute for ambulance chasing."[83] It must be noted that, in the cases we have observed, intervenor lawyers were hardly overpaid and, in fact, often had good reason to worry about where their next check would come from and when it would arrive.

The use of technical interrogators has been suggested and applied in an attempt to make cross-examination more meaningful although there is, by no means, unanimity on the value of such interrogation. Vermont Yankee ASLB Chairman Samuel Jensch, after conditionally accepting Marc Goldsmith as a technical interrogator, noted that often technical interrogators lengthen proceedings because of their inability to carefully frame questions and to avoid repetition and argument.[84] It is clear that the hearing structure and some of the individual Board members are partial to attorney participation. Thus, in a dialogue with Dr. Holcomb concerning Dr. Holcomb's ability to cross-examine witnesses in the Midland case, Chairman Murphy commented that "there are great advantages in having somebody who is a lawyer [performing the cross-examination]."[85] The Midland Board on several occasions expressed dissatisfaction over Dr. Holcomb's conduct of the cross-examination, claiming that it was leading nowhere but

---

82 Telephone interview with Paul Turner, Executive Director, Atomic Industrial Forum.

83 Hearings before the Joint Committee on Atomic Energy, January 23, 1973.

84 Vermont Yankee transcript, p. 1662.

85 Midland transcript, p. 5715.

causing great delay. Perhaps out of frustration, perhaps out of concern that the technological issues be explored, members of the Board made it appear that Dr. Holcomb's lines of questioning were not particularly helpful in extracting scientific or technological information for the record upon which a decision would ultimately be made. Given the goals of the Mapleton Intervenors, the group which Dr. Holcomb was representing, it could be argued that delay was, in fact, his goal.

ATOMIC SAFETY AND LICENSING BOARD

> A presiding officer has the duty to conduct a fair and impartial hearing according to law, to take appropriate action to avoid delay, and to maintain order.[86]

[NOTE: Our comments on the makeup and attitudes of Atomic Safety and Licensing Boards are based largely upon our observations in the three principal cases in this study. There is some evidence that recent changes in the composition of the ASLB panel could have a significant effect on Board attitudes.[87]

It must be noted, however, that various alterations in rules and personnel at the AEC have for years been cited as indications that the approach of the AEC to the licensing process and to citizen participation was changing. Almost certainly the changes will continue. Yet there is no indication that changes made in the past, including the appointment of James Schlesinger as AEC chairman or the apparent willingness of the AEC to abide by the terms of the Calvert

---

86 AEC Rules of Practice, 10 CFR §2.718.

87 These changes include the appointment of experts in the environmental sciences to the panel as well as the appointment of hearing board chairmen with few if any past ties with the Atomic Energy Commission and with a stated commitment to the fair and judicious examination of the issues, regardless of the outcome of that examination.

Cliffs decision, have led to any significant differences in the agency's approach to citizen participation, despite claims that such actions signalled a new relationship between the AEC and the public. Typical of these assurances is the following from the Commission's letter of transmittal of the Annual Report of the Atomic Energy Commission for 1972. Commenting on "the evolving new look of the AEC's regulatory process" the letter notes that

> The Regulatory organization was streamlined to stabilize the licensing situation and to make the process more responsive to the public interest. New leadership was acquired and the staff was strengthened by adding more safety and environmental experts, and new tools for management analysis and control were introduced.][88]

It is apparent from our observations that ASLB's approach licensing hearings with a predisposition to license or issue a construction permit for the plant under consideration. Conversations with several ASLB members indicate quite clearly that the Board members generally view their role as that of presiding officers in hearings to license the plant rather than in hearings deisgned to determine in an impartial manner whether or not to license the plant.[89] The difference is significant; furthermore, it is in no way evident in the AEC's regulations that the Board members' interpretation of their role is proper. The Boards' predisposition to license is probably one of the root causes of

---

[88] U.S. Atomic Energy Commission, 1972 Annual Report to the Congress: Regulatory Activities, p. iii.

[89] It should be noted that it is the stated purpose of Atomic Safety and Licensing Boards to examine whether or not a plant meets AEC standards and regulations and not to make ultimate determinations about the site, the need for nuclear power nor about the comparative benefits of nuclear power and other forms of power generation. This limitation on the power of the Board, while critical to its ability to make decisions on the course of energy policy in this country, ought not limit the ability of hearing Boards to conduct fair and impartial hearings.

their often evident displeasure and annoyance with intervenors and their attorneys. For, if the hearing will inevitably result in a license, then the intervention must be viewed as an essentially delaying procedure and impatience is understandable if unwarranted. (This is not to say that the Boards are not provoked, at times, by the actions of intervenors and their attorneys. In both the ECCS and Midland cases some ASLB reactions can be attributed directly to the conduct of the intervenors' lawyers. But we speak here of the Boards' evident predilection, rather than its response to any given provocation.)

Representative Lester Wolff has noted that

> . . . it seems appropriate to take a long look at the safety and licensing board which is composed entirely of people with a special interest favoring nuclear energy. Any group appearing before the board is made aware very quickly of the overwhelming desire of the board to approve license applications. A more objective board would command greater confidence.[90]

Intervenors are outspoken in their criticism of ASLB makeup. Mrs. William Carl of the Lloyd Harbor Study Group has noted that "staffing the Board with AEC 'experts' only is a travesty of public interest." There is no question that ASLB's have been composed of either AEC employees or individuals with long term relationships with the AEC. In addition, the ASLB's in the three cases observed had no representatives who could be identified with any of the environmental sciences or related fields; recent appointments to the ASLB panel have increased the number of environmental scientists available for appointment to hearing boards.

In February of 1972 the AEC appointed four environmental specialists to its Atomic Safety and Licensing Board Panel. In announcing this response by

---

90 LH, p. 153.

the Commission to the Calvert Cliffs decision, then Chairman Schlesinger asserted that

> We believe that this constitutes another important step in our effort to be responsive to the concern of many individuals and groups over the possible environmental impact of nuclear power plants.[91]

In the cases we have observed, it was apparent that the ASLB could not be considered as an objective body capable of dissociation from the agency to which it is responsible. Efforts at judiciousness are necessarily undermined by the AEC's commitment to the promotion of the growth of nuclear power. (A discussion of the role of the Commission in the nuclear power plant controversy is found elsewhere in this Chapter.)

The question of bias on the part of members of licensing boards has arisen recently in two cases, the Midland and LaSalle County nuclear plant construction permit proceedings. In 1972, ASLB Chairman Arthur Murphy authored an article on the effect of the National Environmental Policy Act on the AEC's licensing process. That article, which draws upon the Midland case for some of its substantiating documentary evidence, appeared in the October, 1972 issue of the Columbia Law Review,[92] before the issuance of the Board's initial decision in the Midland case. The article made the point that

> . . . the job required of the Atomic Energy Commission by NEPA, as interpreted by the courts, is one which the agency cannot perform, and, in any event, one which the licensing process as it currently exists is ill-designed to handle.[93]

---

[91] "Environmental Specialists Named to AEC Public Hearing Boards", AEC Press Release #P-31, February 4, 1972, p. 1.

[92] Arthur W. Murphy, "The National Environmental Policy Act and the Licensing Process: Environmentalist Magna Carta or Agency Coup de Grace?", 72 Columbia Law Review 963, October, 1972.

[93] Ibid., p. 965.

Citing several court decisions, Myron Cherry moved to require that the Presiding Officer recall and revoke the initial decision and to declare the Atomic Sfaety and Licensing Board biased because of Murphy's opinions as stated in the article and because of other statements made by Board members during the course of the hearings. Cherry noted that the test which must be applied in order to determine whether a presiding officer may be disqualified for bias is whether "a disinterested observer may conclude that the agency [or presiding officer or Board] has in some measure adjudged the facts as well as the law of a particular case in advance of hearing it."[94] Thus

> . . . if the presiding officer or the Board in the Midland dockets had in some measure decided in advance that applicant was entitled to construction permits or that a fair and objective NEPA or safety hearing could not be had, then bias is clearly established under the law approved by the Supreme Court of the United States.[95]

Cherry found it "clear upon an examination of the transcript record . . . and . . . of the article written by Arthur W. Murphy" that "the Intervenors faced a Licensing Board infused by a personal bias, thus rendering the initial decision a nullity."[96]

The Appeals Board denied the motion on the grounds that it "was not asserted in a timely manner" but went on to note that "even if it had been timely asserted, there has been presented no basis which would require us to disqualify Chairman

---

[94] Gilligan, Will & Co. v. SEC, 267 F. 2d 461, 469 (2d Cir.) cert denied, 361 U.S. 896 (1959). Quoted in Motion and Supporting Argument to Require the Presiding Officer to Recall and Revoke the Initial Decision and to Declare the Atomic Safety and Licensing Board Biased by Saginaw Intervenors, January 7, 1973, p. 4.

[95] Motion by Saginaw Intervenors, January 7, 1973, p. 6.

[96] Ibid., p. 7.

Murphy. . . ."[97] Prejudgment, it notes, is a ground for disqualification. But the Board draws a distinction between facts and legal issues and notes that "the courts have made it clear that the prejudgment which is proscribed is a prejudgment of facts in controversy but not of legal issues."[98] The Board quotes a case in which a Secretary of Agriculture was not disqualified from a remand proceeding despite his public criticism of the court decision which had ordered the remand:

> That [the Secretary] not merely held but expressed, strong views on matters believed by him to have been in issue, did not unfit him for exercising his duty in subsequent proceedings ordered by this Court.[99]

The Board notes that in a similar case involving the Federal Trade Commission handling of a basing-point pricing system, the Supreme Court held that:

> . . . the fact that the Commission had entertained such views as a result of its prior *ex parte* investigations did not necessarily mean that the minds of its members were irrevocably closed on the subject. . . .[100]

The Board ruled that Chairman Murphy did not prejudge factual, as distinguished from legal or policy, issues:

> It is true, of course, that the article expressed skepticism regarding whether the requirements of NEPA could be fulfilled in the existing licensing process. But the exposition of such thoughts on legal matters was sanctioned by the Supreme Court in its *Morgan* and *Cement Institute* decisions.[101]

----

97 *Memorandum and Order* of Atomic Safety and Licensing Appeal Board (ALAB-101), February 15, 1973, p. 5.

98 *Ibid.*, p. 15.

99 *United States* v. *Morgan*, 313 U.S. 409, 421 (1941), quoted in ALAB-101, p. 15.

100 *Federal Trade Commission* v. *Cement Institute*, 333 U.S. 683, 701 (1948) quoted in ALAB-101, p. 16.

101 ALAB-101, p. 20.

The Board found no evidence of prejudgment of fact:

> All that we find is an individual who may have certain crystallized views . . . on the application of NEPA to the Commission's licensing process. Previous decisions of this Board and the Commission have explicitly recognized this situation as nondisqualifying.[102]

Interestingly, the Atomic Safety and Licensing Appeal Board was forced to rule on a similar claim of bias made by the Commonwealth Edison Company against a technical member of the Board inthe construction permit proceeding on the LaSalle County Nuclear Power Station. In that case, the Board member's comments at a prehearing conference indicated a possible prejudgment of fact concerning the recreational value of the plant's cooling pond. The Appeal Board again drew the distinction between facts and legal issues, and, in its opinion, quoted from a court opinion which disqualified a commissioner of the FTC and which held:

> Once an adjudicator has taken a position apparently inconsistent with an ability to judge the facts fairly, subsequent protestations of open-mindedness on his part cannot restore a presumption of impartiality. . . . We must presume that a fair hearing was denied if a disinterested observer would have reason to believe that the Commissioner had in some measure adjudged the facts . . . of a particular case in advance of hearing it.[103]

The Appeal Board ordered that the ASLB member should be replaced. The Commission subsequently overruled the Appeal Board, arguing that they judged the Board member's remarks "as an effort to elicit additional information and . . . they reflected no more than tentative evaluations."[104] The Commission went on

----

[102] Ibid., p. 22.

[103] Texaco, Inc. v. Federal Trade Commission, 336 F. 2d 754, 764 (D.C. Cir. 1964) vacated and remanded on other grounds, 381 U.S. 739 (1965) quoted in ALAB-102, pp. 11-12.

[104] Memorandum and Order, U.S. AEC, March 2, 1973, p. 3.

to note that

> preliminary assessments, made on the record, during the course of an adjudicatory proceeding--based solely on the decision-maker's judgement to material properly before him in the proceeding--do not compel disqualification as a matter of law.[105]

The Court decisions in the area of what constitutes prejudgment make it clear that a highly subjective determination is required to separate factual prejudgment from prejudgment of legal or policy issues. Further, it is by no means clear that prejudgment of legal or policy issues does not have a significant effect upon the ability of an adjudicator to make an impartial judgment on the facts.[106] Indeed, it is logical to conclude that a predisposition in favor of nuclear power (a "policy/issues prejudgment" that is evident in the attitudes and statements of many ASLB members) is virtually certain to result in a lack of objectivity in resolving factual disputes concerning radiological safety, environmental values, and other points of concern. By the Appeal Board's rulings, such predetermination does not serve to disqualify an adjudicator. We conclude however, that it is precisely the predisposition of Licensing Boards (and the AEC) to grant licenses and permits based upon their preconceived support of nuclear power which constitutes the gravest and most meaningful threat to real citizen participation in the decisionmaking process.

The Midland Appeal Board decision lists several causes for disqualification other than prejudgment of fact of which at least one deserves discussion.

---

[105] Ibid., p. 4.

[106] The relevance of this point is not limited to nuclear power proceedings; questions of prejudgment no doubt pervade all regulatory processes and frequently, we assume, at least some civil and criminal adjudicatory proceedings.

Court rulings have made clear that not only is prejudgment of factual issues a cause for disqualification but that the appearance of bias or prejudgment of factual issues would also suffice to disqualify an adjudicator. That is, apparent bias is a cause for disqualification, just as actual bias is. Thus, the D.C. District Court has held that

> An administrative hearing . . . must be attended, not only with every element of fairness but with the very appearance of complete fairness. Only thus can the tribunal conducting a quasi-adjudicatory proceeding meet the basic requirements of due process.[107]

It is recognized that such a rule may, under certain circumstances, lead to the disqualification of a truly unbiased adjudicator. However, the risk of such an occurrence is less, at least in the eyes of the court (and the Appeal Board) than the risk of a biased hearing. The test of appearance of bias is whether a disinterested observer would have reason to suspect bias on the part of the adjudicator. In the three cases which we have examined in detail, it would be difficult to conclude that the ASLB's (at least at times during the hearings which we have observed) do not display a bias toward the licensing of a plant or in rule making hearings, in favor of industry representatives. Statements and actions on the part of Board members, on the record and off, make it clear that they often view citizen intervention as a purely delaying tactic, a waste of time.

In fact, the very physical set-up of the hearing gives an "appearance of bias." An AEC appointed Board (consisting of members with direct relationship to the AEC and the nuclear development program), the applicant (or in the case

---

[107] Amos Treat & Co. v. SEC, 306 F 2d 260, 267 (D.C. Cir. 1962) quoted in ALAB-101, p. 17.

of ECCS, the vendors and utilities), the AEC Regulatory Staff (which has already determined that the application meets requirements and that it supports the issuing of a permit or license, or promulgation of a rule under discussion, else there would be no hearing in the first place), and perhaps the primary beneficiary of the plant (such as Dow Chemical Company in the Midland case), all with an expressed or apparent interest in the issuance of the permit or license or upholding the rule, present a formidable force against which an intervenor (often represented at a hearing by a single attorney) must attempt to countervail. The audience, as might be expected at a hearing held during working hours, as all we have observed have been, is often made up mostly of those with a direct pecuniary interest in the plant under consideration (i.e., employees of the applicant). (It might be noted that, at one point during the temporary operating license hearings of Vermont Yankee, the composition of the audience was so evident that Chairman Jensch had to ask some employees of Vermont Yankee Nuclear Power Corporation to vacate their places so that the few members of the public who were present could find seats.) The "appearance" of the hearings is that of a proceeding designed to grant a permit or a license and to accord courtesy, patience, forbearance and strict trappings of procedural due process while the intervenor has his day in "court."

Another cause for disqualification of an administrative trier of fact is personal bias against a participant. This is, of course, a difficult thing to prove, but there is no question that a disinterested observer in the Midland and ECCS cases would find it impossible to overlook the very apparent antagonism on the part of some Board members toward some attorneys. This is not to say that the antagonism is not occasionally provoked. Indeed, the behavior of many of the lawyers in those cases (lawyers representing most of the parties present) could be interpreted as provocative although it must be reported that Board hostility clearly was directed, for the most part, toward the intervenors.

Moreover, we have observed on many occasions in the hearings we have attended, physical demonstrations on the part of Board members of annoyance, frustration, pique and anger, physical motions which do not get recorded in the transcript of the hearings. These include headholding, eye shading, eye rolling, looks of disgust, grimacing, tight closing of eyes, etc., indicating clear appearance of annoyance if not outright demonstration of bias.

The nature of the hearing Boards has been a matter of considerable concern to many of the parties involved. Even now, one-quarter of the members of the ASLB panel are full-time employees of the AEC; the others all serve on hearing Boards part-time. This situation leads, quite clearly, to scheduling problems which are particularly acute when hearing Board members combine their ASLB duties with those of university teaching. It might be noted that about one-third of the panel is made up of members of university communities.

ROLE OF CITIZEN GROUPS

We turn next to the role that citizen groups play in the decisionmaking process concerning nuclear power. The question of the citizen's role in making decisions in a highly technological society has been debated often. There are those who hold that decisions about technology and its use should be made by technical experts, those most likely to know and understand the details of the technology. Dixy Lee Ray, recently appointed Chairperson of the Commission, was quoted in the publication The Weekly Energy Report as having said:

> The fact is that in a social decision, when it is taken, any citizen has as valid an opinion as anyone else. But with a scientific and technical decision, this is not true, and an expert's opinion must carry more weight than that of a non-expert. It is a distinction that we must be careful to make.[108]

There are others who argue that questions of the use of technology are social questions, not technical ones at all, and that the decisionmaking process on matters involving the use of technology should be kept out of the hands of the experts.

There was a time, not long ago, when assessments of the effects of technology were performed by experts based upon narrow considerations of technical efficacy and direct, observable results. But our growing awareness of the pervasiveness of the impacts of technology has led to the inclusion, in the assessment process, of the indirect, less tangible, social and environmental effects that often accompany new technology. The recent recognition of the pervasiveness of the impacts of technology has resulted in a skepticism (no doubt healthy) about the ability of "experts" to anticipate all impacts of technology and to make final decisions about its use on behalf of the public. Some citizens have sought, as a result, a greater role in decisions about the adoption of a technology and its use. And the value of citizen participation has been recognized at all levels. In 1971, President Nixon noted:

> In the final analysis, the foundation on which environmental progress rests in our society is a responsible and informed citizenry. My confidence that our Nation will meet its environmental problems in the years ahead is based in large measure on my faith in the continued vigilance of American public

---

108 <u>The Weekly Energy Report</u>, Vol. 1, No. 1, February 12, 1973, p. 7.

> opinion and in the continued vitality of citizen efforts to protect and improve the environment.[109]

Increased public participation is particularly noticeable in the controversy over nuclear power. That at least part of the motivation for citizen involvement is a reaction to the dominance of experts and the powerlessness of citizens is clear in the following excerpt from the testimony of Mary Sinclair (a principal of the Saginaw Intervenors in the Midland case and of the Consolidated National Intervenors in the ECCS case) before the Michigan Public Service Commission:

> . . . [T]here is a myth that only technical experts are capable of forming valid judgments as to scientific and engineering applications in public policy issues.
>
> . . . .
>
> . . . [T]he most compelling argument for the value of citizen participation in the decision making process is the fact that by now it is clear that citizens have for too long left the decisions in energy development to the so-called experts,--to the corporations, commercial organizations, and government agencies who developed powerful vested interest in certain decisions. The fact that these decisions were not adequately assessed by public scrutiny is one of the reasons we find ourselves in deep trouble in many aspects of our economic, social and political development today.[110]

There is no question that underlying most interventions in power plant cases is a profound distrust of experts, of established interests, and of government decisionmakers. The distrust runs so deep that virtually all statements of experts and the "technological establishment" are viewed with great skepticism and almost never accepted at face value. It is this skepticism that has resulted in the failure of public relations efforts to gain public acceptance

----

109 Quoted in La Raza Unida v. Volpe (U.S. District Court, Northern District of California) 4 ERC 1800.

110 Mary Sinclair, "Statement to the Michigan Public Service Commission", (no date).

of nuclear power; it is clear that the mere words of the industry or the AEC hold little or no weight with citizen intervenors who view those organizations as a government-industry complex--a community of vested interests.

Citizen groups involved in nuclear power cases give several reasons for their participation. These tend to fall into three categories: (1) a concern that the plant be made safe, or as safe as possible and that detrimental impacts of its operation be minimized, (2) a concern that public information and education about nuclear power be upgraded, and (3) an attempt to delay or stop the construction and operation of the plant. It should be emphasized that not all of the groups involved cite all of these reasons for their involvement.

The concern for safety and minimizing of impacts of operation is reflected in citizen groups use of hearings to raise such issues as: the effects of radioactive emissions, nuclear safety (including possibility of failure of emergency core cooling systems, fuel densification, hydriding, the effects of the total fuel cycle, environmental effects, etc. Some argument can be made that it is the concern of citizen groups that led to the raising and airing of these issues and that without their participation the issues would have been glossed over or, as far as the public is concerned, ignored. Thus, Arthur Murphy, in his controversial Columbia Law Review article noted that:

> It may be true that the public will not often add much to the information supplied by experts, but if nothing more, the activities of environmentalists have provoked a comprehensive reexamination of the issues.[111]

Surely, citizen groups and their attorneys feel that this is so. Myron Cherry

---

[111] Arthur Murphy, "The National Environmental Policy Act . . .", p. 1007.

has declared that "I and others have performed a public service in making sure that the ECCS issue is fully debated."[112] In another forum, Cherry commented:

> The environmentalists who have opposed nuclear power plants on safety and environmental groups [sic] have not been irresponsible. They have positively contributed to a public discussion of the problems facing the nuclear industry.
> . . . .
> The information which has now surfaced as a result of their efforts has confirmed that these organizations have been right and . . . that without their efforts and without public hearings, the AEC and the nuclear industry would have been successful in hiding their "bad news."[113]

Others agree that the public hearing process has yielded significant benefits. Governor Wendell Anderson of Minnesota noted that ". . . many of the most serious and searching questions have been posed to the AEC as a result of the public hearing process.[114] Still others are less certain that the citizen participation has been helpful. Glenn Seaborg, former AEC Chairman, for example, feels that "there has been a deliberate attempt to accent the fearful, the scary, the derogatory."[115] Joint Committee on Atomic Energy member Craig Hosmer has stated that AEC hearings have "increasingly been used by mischievous nuclear opponents to blackmail the utilities and the Commission into giving them their way."[116]

---

112 Letter from Myron Cherry to Arthur Murphy, April 15, 1972.

113 H.R. 13731 and H.R. 13732 to Amend the Atomic Energy Act of 1954 Regarding the Licensing of Nuclear Facilities, Hearings before the Joint Committee on Atomic Energy, March 16-17, 1972, p. 213. (Hereafter cited as Licensing Amendment Hearings.)

114 Ibid., p. 447.

115 LH, p. 22.

116 Quoted in the National Journal, June 17, 1972, p. 1027.

Most of the citizen groups involved in the cases we examined expressed a goal of "public education." It is, however, by no means clear just what was meant by "public education." To some, the phrase meant giving the public information about nuclear power plants with no real goal in mind; to others it meant frightening the public; to still others it meant attempting to organize the public in order to strengthen support for the group's other goals. One can, of course, raise the question of whether the AEC's hearings provide a viable forum for public education, however it is defined. For the fact is that the AEC's hearings are poorly attended and generally, poorly reported in the media which is a reflection of the fact that to the extent the hearings deal with technical facts, they are, as we have noted, to most people, boring and incomprehensible. Information introduced in the hearings is rarely presented to the public in a manner designed to provide enlightenment about nuclear power, its benefits and its dangers. James Campbell, President of Consumers Power Company, has pointed out the difficulty of using adversary proceedings to "educate." Although his remarks are particularly critical of citizen intervenors, they can, we think, be applied just as well, to all of those taking part in AEC hearings:

> Mandatory AEC license hearings have been justified as desirable to educate the public. Experience has demonstrated, however, that members of the public rarely attend uncontested hearings, and that members of the public who attend contested hearings are usually allied with one of the parties and are less interested in being educated than in seeing a vindication of their preconceived ideas about the case. Education would be better confined to more conventional means such as school programs, public information forums, science fairs and the like.[117]

----

117 LH, p. 440.

In the three cases we have examined, only one group, the Mapleton Intervenors at Midland, did not cite public education as a goal, and in fact, cited cancellation of the project as its primary goal. Wendell Marshall, a member of that group, stated, "We don't want it [the plant] even if it's good."[118] More recently Mr. Marshall filed suit in a Midland Circuit Court to obtain an injunction against plant construction. His comment at that time was "In Mapleton we're staked out for the kill, like an Indian over an ant hill."[119] But this is an extreme view. To be sure, the fact of citizen intervention leads to more questions being asked, more documents being produced, more parties being accommodated in the scheduling of hearings. But the line of demarcation between legitimate debate and intentional delay is often hazy. Arthur Murphy has noted that

> . . . the types of questions involved in these proceedings--and in any power plant hearing on environmental questions--are such that the potential for inquiry is virtually limitless.[120]

Despite the fact that delay is not often the primary goal of citizen groups intervening in licensing hearings, it is often this aspect of interventions that is strongly criticized by the AEC, the nuclear industry, and other proponents of nuclear power. George Trowbridge has commented that

> . . . we have seen . . . the ability of intervenors representing small segments of the public, however sincere their purpose, to use trial techniques and procedures to hold up indefinitely, the issuance of licenses, either in the interest of defeating the project or of forcing settlements

---

118 Interview with Wendell Marshall, May 19, 1972.

119 Quoted in INFO, Press Edition, No. 29, February 1973, p. 4.

120 Arthur Murphy, "The National Environmental Policy Act . . .", p. 994.

> according to their own notions of nuclear safety or environmental protection.[121]

One of the most significant problems facing intervenor groups is the difficulty of obtaining adequate funds to take part in licensing proceedings. Most local citizen groups depend upon small contributions from local residents, yet the cost of an intervention can run to nearly $100,000.[122] Along with the general unavailability of expert technical help (to be discussed in the next section), the funding problem is perhaps the most important determinant of the adequacy of an intervention. Decisions about strategy and tactics, about possible appeals and court action, all are made with one eye on the budget. For example, although stipulating most factual matters in the Maine Yankee case was part of Harold Green's overall strategy in that case, there is no question that the desperate financial straits of his clients dictated, at least in part, the choice of strategy.

The argument is often made that drawing out a hearing hurts the utility be delaying the construction and operation of the plant in question. While this is no doubt an effect of delay, it is our observation that delay causes problems for intervenors as well. Angus MacBeth of the Natural Resources Defense Council points out that "the situation is particularly difficult for environmental groups which do not have the funds, manpower or technical expertise to expend on repeated skirmishes in a never-ending war." He goes on to note that

> The clients of the environmental lawyer are strapped for funds to fight specific battles and to hire the expertise which will

---

121 LH, p. 334

122 Interview with Diana Sidebotham, NECNP.

> allow them to fight intelligently. The costs of their efforts may from time to time be deductible as charitable contributions, but unlike the utilities, they cannot be written off as ordinary and necessary business expenses or passed along to the consumer as part of the rate base.[123]

The mounting costs of a prolonged proceeding exacerbate the financial difficulties of the intervenor. The result is that hearings are often reduced to battles of attrition between intervenors and utilities with the party with the sparsest resources giving in.

Lack of adequate funds is one of the primary causes for the seemingly unending requests of intervenors for more time to meet procedural requirements. There is usually no delaying motive involved in such requests; the cause is often the difficulty in logistics (research, typing, duplicating, distributing) of meeting deadlines with inadequate funds. The plight of intervenors has led at least one prominent attorney for utility groups (who for obvious reasons has asked to remain anonymous) to propose that applicants for permits and licenses be assessed a fee (perhaps $100,000) to pay the costs of a citizen intervention. The motive is not entirely altruistic; provision of adequate funds to citizen groups would, in this attorney's theory, eliminate the necessity for procedural delay. Presiding officers could be less lenient with intervenors by insisting that deadlines be met. Citizen groups could no longer plead lack of funds as a reason for incomplete or unsubstantial presentations.

Providing fees to underwrite citizen interventions has been a matter of considerable discussion in the courts. The legal system of the U.S. has traditionally refused to award attorney's fees to successful litigants who request

---

[123] Angus MacBeth, Mimeographed paper, June 19, 1972.

them in judicial or administrative actions. To be sure, this is not a hard and fast rule; a number of exceptions have been carved out by the Congress and the Federal courts.

For example, a limited number of Federal statutes provide for mandatory shifting of legal fees where the successful plaintiff has been injured by a party in violation of the state, or an order of an administrative agency. To date, most of these statutory exceptions are restricted to civil rights legislation. For example, the Civil Rights Act of 1964 provides that

> In any action commenced pursuant to this subchapter, the court, in its discretion, may allow the prevailing party other than the U.S., a reasonable attorney's fee as part of the costs, and the U.S. shall be liable for costs the same as a private person.[124]

Other Federal statutes permit fee shifting to either plaintiff or defendant at the discretion of the court. This has been generally limited to situations where one party's acting in bad faith has triggered a fee transfer to be approved by the court. A third category of exceptions consists of a number of extra-statutory, discretionary categories developed by the Federal courts under color of their historic equity jurisdiction.[125]

The legal system has responded to this dilemma in ways which have not proven adequate. Contingency fees, which are based on a percentage of recovery, are often charged by lawyers in cases where they seek monetary damages. However, as prominent environmental lawyer Bernard Cohen put it, "I could have a

---

[124] 42 USC §2000 a-3(b).

[125] "Allocation of Attorneys' Fees" after Mills v. Electric Auto-Lite Co., 38 Chicago Law Review 316, Winter, 1971.

lucrative practice if I only took cases involving monetary damages, but for every one case involving these damages, I take 10 going for injunctive relief."[126]

Faced with a serious problem of clients without attorneys, the courts have begun to broaden the criteria used in determining whether or not attorneys' fees should be awarded the successful litigant. This can be largely attributed to the increasing acceptance of the private attorney-general concept of enforcing legislation.[127] It has become evident that public interest groups or similar plaintiffs seeking to enforce legislation or otherwise affect public policy cannot be successful without adequate representation by counsel.

In Newman v. Piggie Park Enterprises[128] the Supreme Court awarded attorneys' fees under Title II of the 1964 Civil Rights Act to a successful plaintiff in a class action suit to enjoin racial discrimination at drive-in restaurants in South Carolina. The court said, in part,

> When a plaintiff brings an action under that title, he cannot recover damages. If he obtains an injunction, he does so not for himself alone but also as a 'private attorney general.' If successful plaintiffs were routinely forced to bear their own attorneys' fees, few aggrieved parties would be in a position to advance the public interest by invoking the injunctive powers of the federal courts.

In a subsequent attorneys' fee decision, Sims v. Amos,[129] where plaintiffs in an apportionment suit against state officials were awarded attorneys' fees,

---

[126] "Ecology: Win Some, Lose Some", Washington Post, February 8, 1973, p. B1.

[127] A private attorney-general is a member of the public who brings a civil suit in order to secure compliance with a statute or with a broad constitutional right.

[128] 390 U.S. 400.

[129] 340 F. Supp. 691.

the court determined that

> Each case must be decided according to its own peculiar facts. A finding of bad faith is not always a prerequisite to the taxing of attorneys' fees against defendants. . . . Under such circumstances, the award loses much of its discretionary character and becomes a part of the effective remedy a court should fashion to encourage public-minded suits, and to carry out congressional policy.[130]

Obviously, the hard-and-fast opposition to awarding attorneys' fees in the courts is breaking down. In most instances, however, to successfully recover these fees plaintiffs have found it necessary to demonstrate that an applicable statute specifically provides for recovery of the fees under circumstances similar to those of the plaintiff class.

It is in this context that La Raza Unida v. Volpe[131] arose. In this case a private citizens group brought a successful suit to halt Federal and State highway agencies' construction of a Federally aided highway through a park. In a subsequent decision, a U.S. District Court in California awarded attorneys' fees and expert witness fees to plaintiffs. The court found that

> While the general American rule is that attorneys' fees are not ordinarily recoverable as costs absent an express statutory authorization, the courts have developed exceptions to this rule for situations in which overriding considerations indicate the need for such a recovery.[132]

The Court explained that its power to grant attorneys' fees springs from the equitable powers of the court, and are not dependent upon Congressional authorization. Three tests were held to be met by the plaintiff class which

----

130 Ibid., p. 694 (emphasis added).

131 4 ERC 1797.

132 Ibid.

permitted it to exercise its equitable jurisdiction: (1) the plaintiffs' action had effectuated strong public policies promulgated by various environmental legislation, (2) numerous persons had benefitted from plaintiffs' litigation success, and (3) only a private party could have been expected to bring this action.

The La Raza Unida decision presents an excellent discussion of the need for provision of funds to citizen groups which is equally applicable to nuclear power cases. Judge Peckham of the U.S. District Court, Northern District of California, in discussing the decision to provide attorney's fees and expert witness fees comments:

> Responsible representatives of the public should be encouraged to sue, particularly where government entities are involved as defendants. . . . [O]nly private citizens can be expected to "guard the guardians."
>
> However, these exhortations towards citizen participation can sound somewhat hollow against the background of the economic realities of vigorous litigation. In many "public interest" cases only injunctive relief is sought, and the average attorney or litigant must hesitate, if not shudder, at the thought of "taking on" an entity such as the California Department of Highways, with no prospect of financial compensation for the efforts and expenses rendered. The expense of litigation in such a case poses a formidable, if not insurmountable, obstacle.
>
> The only public entities that might have brought suit in this case were named as defendants in this action and vigorously opposed plaintiff's contentions. Only a private party could have been expected to bring this litigation, and yet a private party is least able to bear the tremendous economic burdens. To force the private litigants to bear their own costs here would be tantamount to a penalty, and it seems somewhat inequitable to punish litigants who have policed those charged with implementing and following Congressional mandates.[133]

---

133 Ibid., p. 1801.

In an overall perspective it must be recognized that La Raza Unida does not reflect prevailing legal views; most courts still will not award attorneys' fees without an explicit legislative mandate. It is encouraging to note that more and more frequently these equitable awards are being given in order to encourage the private attorney-general concept of public interest law.

How does this apply to intervention in administrative hearings? Certainly the need for financial support or reimbursement to such an intervenor is just as great as it is to a party bringing a public interest suit in court. David Sive, plaintiff attorney in the famous Scenic Hudson case, points out that

> The ad hoc citizens group or regional or national conservation organization, the typical plaintiff (intervenor) has little money and no paid legal staff. Every product and service that goes into a substantial litigation--lawyers, typists, photostats, expert witnesses--must generally be wholly or substantially donated. The opposing administrative agency, often joined by a large corporation which wishes to appropriate a natural resource, has virtually unlimited or at least ample means.[134]

In the context of administrative interventions at the AEC, the interests in encouraging legitimate meaningful participation by the public in the decisionmaking process leads to the inescapable conclusion that qualified intervenor groups should be awarded legal fees and expert witness fees.[135] It is important to remember that, in court cases, only successful plaintiffs are normally reimbursed. This standard is clearly inapplicable in the context of an AEC licensing hearing to determine whether or not an intervenor group

---

[134] David Sive, "Some Thoughts of an Environmental Lawyer in the Wilderness of Administrative Law", 70 Columbia Law Review 618 (1970).

[135] We realize that a definition of "qualified intervenor groups" is by no means easy to arrive at. It should be noted, however, that 10 CFR §2.714 does spell out what constitutes sufficient grounds for the granting of leave to intervene. We would assume though, that such standards would need to be revised were the AEC called upon to pay for intervenor's participation.

should receive fees. What is a successful AEC intervention? What would success be for the New England Coalition? Or what would constitute success on the part of CNI in the ECCS hearings? Should attorneys' fees be awarded in a rule making procedure where intervenors provided the impetus for the hearings in the first place or where intervenors caused a change to be made on criteria or processes? Could sufficient standards be adopted on such a narrow interpretation?

These sorts of questions need to be dealt with by the AEC and the Congress. An amendment to the Atomic Energy Act might be adopted which would require the AEC to compensate qualified intervenor groups (without an adequate financial base) for reasonable attorney and expert witness fees incurred as a result of their participation in either a rule making or adjudicatory hearing. To determine who is so qualified, a standard might be based on the <u>La Raza Unida</u> criteria concerning the effectuation of strong public policies; in other words, the intervening group should not be forced to prove that it fulfilled all the objectives of its intervention in order to recover the aforementioned expenses. Perhaps successful fulfillment of existing criteria for standing in a proceeding under the Atomic Energy Act of 1954 or under the Administrative Procedures Act might be an adequate basis for such judgment. Denial of standing could then be appealed to the courts by the potential intervenors, costs of which could be reimbursable upon a favorable finding for plaintiff.

Such legislation would provide the courts with legal ammunition for its arsenal of equitable arguments in favor of providing attorney and expert witness fees. Intervenors in AEC hearings should not have to depend solely on the graciousness or generosity of the courts to determine that they are appropriate recipients of equitable compensation for their participation in public policy-making. In the words of the late Justice Hugo Black,

> The courts are interpretors, not creators, of legal rights to recover and if there is a need for recovery of attorneys' fees to effectuate the policies of the Act here involved, the need should, in my judgment, be met by Congress, not by this Court.[136]

One recurrent criticism of citizen groups is that they are accountable to no-one other than their own members and that because of this they tend to take extreme and irrational positions without having to answer to any broader constituency. The criticism holds that such groups, operating under the guise of "public interest" representatives, take irresponsible stands which cause the "real" public to suffer through unnecessary delay and through the propagation of untrue and often frightening misinformation about nuclear power. For instance, Chet Holifield (D-Cal.) of the Joint Committee on Atomic Energy has complained that:

> Any irresponsible, technically untrained person can go before these hearings and make a wild statement and the press will carry it and completely ignore a trained technician who offers a complete refutation of that wild statement.[137]

On another occasion he commented that, in one case, "as I read it, these intervenors are setting themselves up as a judge and jury over and above the Federal authorities and the State authorities";[138] John Quale, President of Wisconsin Electric Power Company, was in agreement with him.

In a hearing on the National Environmental Policy Act in 1972, William Kennedy, Associate General Counsel of the General Electric Company saw the

---

136 Mills v. Electric Auto-Lite Co., 390 U.S. 397.

137 LH, p. 21.

138 LH, p. 251.

effect of NEPA, the Atomic Energy Act, and the Administrative Procedures Act, as interpreted by the courts, as placing "an extraordinary degree of power . . . in the hands of private persons who have no accountability to the general public."[139] AEC Commissioner Ramey has taken a similar approach:

> It is now possible for counsel for a determined intervenor to use delaying tactics, particularly at the operating stage, to obtain concessions from the utilities. These concessions are often in the form of additional "environmental safeguard devices" which may not actually improve environmental protection. Such tactics, pursued in the name of environmental protection, will inevitably pose an extra financial burden on the consumer.
>
> The effect of "nuclear blackmail" is to shift decision-making responsibility to groups relatively limited in size and bearing no public accountability.[140]

One could of course, find numerous similar comments, all with a similar general thrust that intervenors, posing as representatives of the public interest, operate in an irresponsible manner without accountability to the public at large. A response to this argument holds that any party in a contested proceeding views its role as representing some version of the "public interest." Utilities and reactor manufacturers, for example, see themselves as providing adequate electricity for an improved and improving style of life and therefore, acting in the "public interest." The Atomic Energy Commission sees itself as regulating nuclear power to protect the health and safety of the public as well as promoting the public will as expressed by the Congress by supporting the development of nuclear power as a major energy source. In the same way,

---

139 National Environmental Policy Act, Hearings before the Senate Committee on Public Works and the Senate Committee on Interior and Insular Affairs, March, 1972, pp. 469-470.

140 "Nuclear Plant Licensing: Causes and Possible Solutions", in Public Utilities Fortnightly, March 30, 1972.

citizen intervenors often see themselves as protecting some sort of "public interest" against the onslaught of an alliance of industry and government. When pressed, however, representatives of all of the parties admit that the concept of the public interest is an elusive one and that, in reality, any group represents only <u>its</u> interest and perhaps <u>its</u> view of the public interest. It seems unfair to criticize citizen intervenors, for example, for not representing a "real public interest" without recognizing explicitly that the utilities and vendors do not represent that interest either. (The case of the AEC is more complex and is discussed in some detail later.) Intervenors are a party to an adversary process. Their goals vary from those of the Mapleton Intervenors at Midland (essentially: stop the plant at any cost) to those of the Natural Resources Defense Council at Vermont Yankee (essentially: to raise certain issues such as transportation and waste disposal which might not otherwise have been considered). They should be judged on their goals (and how they go about achieving them) rather than on whether or not they adequately represent the elusive public interest. As to accountability, intervenors are responsible to their own memberships in much the same way that utilities and manufacturers are responsible to their owners and stockholders. The representatives of intervenor groups taking part in hearings can no more afford to lose the confidence and support of the groups' membership and supporters than can the representatives of utilities and vendors afford to lose the confidence and support of management and stockholders. In this sense, the representatives of citizen groups are responsible to other parties. They must exercise this responsibility within the rules of the proceeding within which they operate.

SCIENCE, SCIENTISTS, AND CITIZEN GROUPS

For several reasons, the scientific and technical activity of citizen groups in the nuclear power controversy is less than impressive. One could argue that scientific and technical input to the process is at best secondary to questions of policy, procedure, and governance.

But it is clear that decisions about the use of nuclear power, or at least one input to those decisions, contain a scientific and technical component. Applicants for licenses, and the AEC itself, recognize this; they are represented at hearings by an array of scientific and technological expertise and talent. Not so citizen groups. Their scientific and engineering advice often comes from graduate students or from individual and isolated experts on some aspect of science or technology, who frequently have little if any experience or expertise directly related to nuclear plants and their impacts. Quality of such input is a difficult thing to judge in an objective manner, but certainly the impression one gains from observation of licensing and rule making hearings is that, with some exceptions, the level of scientific and technical advice to intervenor groups is marginally adequate, if that. This fact is clearly recognized by the attorneys for the intervenors and often by the scientists and engineers themselves. A principal science advisor in one of the cases under consideration admitted candidly that he did not perform cross-examination because he "didn't feel qualified . . . to do a good job of questioning."[141]

Some of the technical advisors to groups opposing permits or licenses for power plants are proponents of nuclear power and admit their predisposition

---

[141] Interview with a citizen group's technical advisor.

quite frankly. They see their role as playing devil's advocate in an attempt to demonstrate that all is well with nuclear power and the reactor development program. (It is of interest that involvement in licensing proceedings has, in some cases, created a degree of skepticism in these individuals which they admit they did not feel at the outset of their participation.) Some intervenor attorneys find such arrangements less than satisfactory since the adversary proceeding in which they are involved requires them to make a forceful case on behalf of their client's side of the issue. The promotional bias of some of the opponent intervenor's advisors may result in their overlooking some important points at issue, as might any personal ambitions for a future career in nuclear energy.

One result of the inadequacy of scientific and technological advice to intervenors has been the reliance of intervenors, particularly in individual plant hearings, on penetrating cross-examination to build their cases. (This is, of course, not always the case; in the ECCS proceeding the Consolidated National Intervenors' case was made up of both cross-examination and a detailed direct case.) Some attorneys have rationalized this strategy by arguing that it is not the role of intervenors to demonstrate problems with a particular plant or rule but rather it is the role of applicants and vendors to prove the absence of problems (which relate to the burden of proof argument discussed earlier). Nonetheless, all attorneys and citizen groups openly admit that they would prefer to make a strong direct case. Some go so far as to state that it is impossible to make a viable scientific or technological case (one which will significantly affect the decision) by cross-examination alone and that a strong direct case is an absolute necessity.[142]

----

[142] Interview with Ken McCue,

Why, then, is the scientific and technological help obtained by citizen groups usually so inadequate? One clear reason is that such help costs money and virtually all citizen groups lack adequate financial resources. It is true that some experts donate their time, while others aid citizen groups for a small consulting fee. But most experts are just too expensive and others simply too busy doing their own research. This accounts for the widespread use of, and reliance on graduate students for scientific and technological assistance.

Another reason, and one which is constantly alluded to by citizen groups, is the difficulty of finding experts on nuclear power who do not have close ties with the nuclear power industry or the AEC. Dr. Philip Handler, President of the National Academy of Sciences, has written:

> It is essentially impossible to find more than a handful of . . . experts who have not, in relatively recent times, had significant support--either research grants or actual employment--from the U.S. Atomic Energy Commission.[143]

Such experts, out of concern (real or imagined) for their positions and reputations, find it difficult to support or testify on behalf of opponents of nuclear power plants. Others simply do not care to expose themselves to the heat of the adversary arena; do not care to be embarrassed or made to look bad by clever lawyers. They are aware of the limited and discrete nature of their expertise, have come to no final conclusions in their work or prefer to do their "testifying" in academic journals. Often they feel that testifying on behalf of opponents of nuclear power plants is not likely to further their own career or to win them the acceptance of their peers. This leaves the citizen

----

143 H. Peter Metzger, The Atomic Establishment (New York: Simon and Schuster, 1972) p. 26.

group in the position of trying to find expert assistance in the face of great reluctance on the part of virtually all of the acknowledged "experts" in the field to help. This too leads to the reliance on graduate students, or on fortuitous or unique circumstances. Examples of the latter include the phenomenon of Daniel Ford, a bright young man with time, inclination, and ability to become an expert on emergency core cooling systems; Henry Kendall, a distinguished high energy nuclear physicist with the inclination and ability to understand nuclear power plant operations; or John Gofman and Arthur Tamplin, two scientists with the courage to speak out about their concerns. Harold Green, often an attorney for citizen intervenors, has pointed out that intervenors are often "forced to present witnesses who are far outside the establishment in terms of their competency and experience in the area."[144] Myron Cherry emphasizes the difficulty of obtaining expert witnesses or assistance:

> I have talked to academicians at some of the leading institutions in the United States. I have asked them to support me objectively and fairly in just analyzing information that I have found and that I want to bring to the attention of the board. Every one of them is afraid to testify because of relationships they have with the Atomic Energy Commission. That is what is going on in the field. . . . There is subtle intimidation, the academicians are being purchased one way or another through funding that is coming to the schools. . . . The industries are funding efforts and too many consultants are too afraid to take on the hand that is feeding them.[145]

Irving Like has also commented on the intimidation of experts; he notes:

> . . . the fear or reluctance of qualified scientists and technicians to testify against a project recommended by the Atomic Energy Commission. Various factors have dissuaded scientists and engineers (even those with strong

---

144 LH, p. 356.

145 LH, p. 422.

> concerns about the environmental effects of nuclear power projects) from testifying in behalf of citizens in such litigations. These include loss of employment or consultant contracts with the AEC or the utilities and the industry engaged in the nuclear power technology, curtailment of research grants to universities with whom the prospective witness is affiliated, and other forms of economic and professional harassment.[146]

We have looked carefully for evidence of any intimidation or harassment of technical experts. At Midland, there has apparently been some punitive action taken against Dow engineers who supported the intervenors (in the form of arbitrary transfers or temporary assignments to other locations, for instance). But the intimidation is usually quite subtle. Of two Dow engineers who indicated the existence of harassment by the company, only one stated, without question, that his treatment was due to his involvement in opposing the nuclear plant. The other suspected his ties with the intervenors as a cause, but would not categorically state that he was being punished for his anti-plant activities. One of the initiators and principal science advisors to the New England Coalition, Irving Lyon, has been dismissed from his position at Bennington College, and Lyon feels that his dismissal was directly connected with his involvement in the Vermont Yankee case. (The President of Bennington, at that time, was a former AEC attorney.) There are others, however, who attribute the Lyon dismissal to ordinary intra-university politics and who claim that Lyon's Vermont Yankee activities had no bearing whatsoever on the dismissal. We do not have enough evidence to agree or disagree with either of these opinions. (Suffice to say, the case of Lyon's dismissal is a strange and complex one. No reasons for the dismissal--actually, the non-renewal of a contract--have been given.

----

146 Like, "No-Win Strategy" at LH, p. 1403.

Ironically, Harvey Carter, a lawyer representing the Conservation Society of Southern Vermont in its intervention at Vermont Yankee, also represents Bennington College in some matters, one of which is the case of Irving Lyon.) At the ECCS hearings, Henry Kendall, who has for years received AEC grants for research in high energy physics, did not encounter any overt attempts at intimidation, but does recognize the more subtle displeasure expressed by some colleagues at the Massachusetts Institute of Technology. On the other hand, John Gofman and Arthur Tamplin, who published the results of their research on the relationship between radiation exposure and cancer while employees of the Lawrence Radiation Laboratory in Livermore, California, were apparently subjected to overt attempts at AEC censorship and to other forms of what they describe as intimidation and harassment by the AEC.[147] Gofman, widely recognized for his eminence as a biomedical researcher, has since left the AEC funded National Laboratory following the severe reduction of funds to support his research and that of his scientific staff.

But harassment aside, it is evident that there is a paucity of available nuclear engineers and physicists who are willing, publicly at least, to identify themselves with intervenors. Government funding for nuclear reactor research comes, for the most part, from a single source. The exclusivity of AEC sponsorship of reactor research and development is a basic and perhaps insurmountable obstacle to a free and open debate within the scientific community and therefore, in the public arena, on matters involving the safety and other impacts of nuclear technology. (In areas of science, where funding for research and

----

[147] Interview with John Gofman, May 16, 1973.

development can be obtained from multiple sources, intervenors have less trouble obtaining scientific assistance and are limited primarily by their own financial resources. This accounts for the fact that intervenors can obtain the services of experts on fish biology, meteorology, noise and other environmental issues.) The AEC's role as the major funding source for research in the nuclear sciences over the last 25 years has created a situation in which almost every nuclear scientist and engineer of worth is under grant or contract to AEC, or in the employ of the AEC, the nuclear industry, or a national laboratory. This is not to suggest that there is absolutely no dissent in the nuclear community. Gofman and Tamplin, though clearly exceptions to the rule, indicate that despite long standing connections with the AEC, scientists (at least those with a degree of prominence in their field) can express their independence of the AEC and the nuclear community, though not without a willingness to be subjected to sanctions. More often, however, expressions of dissent from within the nuclear community is manifested by the anonymous transmittal of AEC internal documents sent through the mails to intervenor attorneys. Much of the crucial data underlying the case presented by the Consolidated National Intervenors in the ECCS proceeding was acquired in this manner.

A third reason for the inadequacy of the scientific and technical activity of intervenor groups is a lack of inclination, at times, to obtain the best available scientific and technical input. Some intervenors feel, and often rightly so, that procedural matters are often of greater import to their position than are scientific and technical matters. Since a principal reason for interventions is, as has been noted, a concern with governance and with the role of the citizen in the governmental decisionmaking process, intervenors are often concerned more with being permitted to take part in an intervention than with technical details. Another point should be noted here: scientific and technical details of nuclear power plants are not interesting or exciting to anyone, with the possible exception of nuclear scientists and engineers. In their role as "public educators" some intervenors see the necessity of making their participation exciting and newsworthy, and, therefore, noticeable to the public. (This is the principal thrust of Irving Like's paper, "Multi-Media Confrontation--The Environmentalists' Strategy for a 'No-Win' Agency Proceeding.") There is no question that rule making and licensing hearings lose any excitement which they might have when the discussion turns to technical details.

Last June (1972), for example, a State Siting Committee meeting on a strongly opposed nuclear plant for the small town of Seabrook, New Hampshire, which drew a full-house crowd in excess of some 350 persons to the local high school auditorium. The hearing convened at 9:00 am on a mid-week workday which was marked by exceedingly heavy rain. The president of the applicant utility company was leadoff witness. His presentation delivered in a barely audible monotone droned out economic projections and a detailed forecast of New England power needs for the next 50 years, or so. During the latter part of the morning a scan of the audience revealed a number of citizens who, though exercised at the prospect of a plant in their midst, were bored to sleep. After a break for lunch, the audience had shrunk to no more than 50 people. At least two utility's attorneys, to whom we subsequently related our observation, averred that it has become standard procedure at such hearings where large numbers of citizen turn out to "cool them down, by highly technical or statistical presentations." It can, of course, be validly argued that such is the true nature of the matters at issue and that there is nothing particularly "sexy" or exciting about nuclear technology or electric generation or statistical projections of power consumption.

It is, moreover, evident that what public interest there has been in the ECCS hearings, a forum where a major issue of nuclear safety is being discussed, has been the result of the legal and procedural wrangling rather than the technological details of the adequacy or inadequacy of core cooling systems.

For the most part, witnesses and scientific advisors for intervenor groups act as individuals rather than as interdisciplinary or multi-disciplinary teams. This is dictated in large measure by the unavailability of both funds and experts, but it works great hardship upon the intervenors and often makes it impossible for them to present a coherent scientific and technological case. By contrast, applicants and the AEC make use of teams of experts to develop and present their cases. They recognize that matters involving the application of technology and an understanding of the effects of the use of the technology are essentially interdisciplinary in nature. The proper study of such matters must necessarily be performed by groups of representative scientists in areas relevant to the problem at hand. Such a group must work as a team and adhere to demanding requirements for interactive study and communications, which requires long and intensive periods of time during which they need to work together. But the kinds of scientific and technical assistance which citizen intervenors get and use is characterized by specialized and particularized knowledge. Individual witnesses and advisors find it understandably difficult to deal with the breadth and depth of the problems raised. Uncomfortable outside their own fields of interest, intervenor witnesses and advisors find it impossible, as individuals, to consider the complex interrelationships involved in assessing the safety, environmental, and other impacts of nuclear power plants.

Technical interrogators have been used by intervenors, at times, in each of the three cases under consideration. The general attitude of hearing Board members seems to be that it is usually best to have lawyers ask questions of the witnesses, but exceptions are made. Intervenor attorneys have requested permission to employ experts as technical interrogators when they felt that they themselves could not adequately question certain scientific and technical experts. The expert interrogator, they argued, was needed to expedite the hearing process and to facilitate obtaining more meaningful information from witnesses. On occasion, the use of technical interrogators has served precisely this purpose; the use of Daniel Ford as an interrogator during the ECCS hearing most certainly led to a more meaningful cross-examination of vendor and AEC witnesses than would have been possible had Myron Cherry handled all of the questioning. But at times, the use of techncial interrogators has had the opposite effect. For instance, in the Midland case, technical interrogator Dr. Lawrence Holcomb often reacted more as a witness and advocate than as an interrogator. His argumentative questions, his dwelling for long periods on obviously insignificant details, and his difficulty in phrasing questions and in making salient points, often angered the hearing Board and the attorneys for the applicant, the AEC, and Dow Chemical Company.

One difficulty in the use of technical interrogators is the requirement in AEC regulations that technical interrogation "shall be limited to areas within the expertise of the individual conducting the examination or cross-examination."[148]

---

[148] 10 CFR §2.733.

Thus, technical interrogation is limited to relatively narrow areas since, as indicated earlier, single individuals generally possess expert knowledge in only a limited area. Virtually any line of questioning will expand the scope of inquiry beyond the expertise of the questioner and puts the hearing Board in the difficult position of either suspending the questioning or permitting the limits set in 10 CFR §2.7333 to be stretched. The first course invariably incurs the objections of the attorney who requested the technical interrogator; the latter often leads to the posing of irrelevant or undirected questions. This undermines the purpose of permitting the technical interrogator in the first place, which is to expedite and facilitate obtaining meaningful information, thus making a record upon which a decision can be made.

Of great importance (perhaps far greater than necessary) in nuclear power proceedings is the definition of "expert" and the determination of who is an expert. The AEC Rules of Practice seems to define an expert as "a qualified individual who has scientific or technical training or experience."[149] Nuclear physicist Ralph Lapp has a simpler definition; he defines an expert as "a member of the club." The "club" here is meant to include other scientific and technical experts. Experts are permitted to act as technical interrogators in hearings and play a different role than laymen in the hearings. Not the least of the differences between the participation of experts and that of laymen is the manner in which they are treated by the attorneys as well as by some members of the Board. Further, it is apparent that the greater the status of an expert within the scientific community, the more deferential is the treatment

----

[149] 10 CFR §2.733.

of the expert by the Board, particularly the technical members, and the greater the latitude afforded him. This observation lends great credibility to the Lapp definition of an expert.

The fact that ASLB members and attorneys tend to take a rather traditional view of the expert, leads to some interesting phenomena. Henry Kendall, a high energy nuclear physicist, is accepted by the ECCS Board as an expert because of his Ph.D. and his description as a physicist. Daniel Ford, trained in economics, is not accepted as an expert because of his lack of a Ph.D. or scientific publications. Yet neither is trained or experienced in emergency core cooling system design or operation. Both gave the matter of core cooling systems great attention and study; both became highly knowledgeable in this particular area of nuclear technology. Both are men of considerable intellectual qualifications and both made inputs to the ECCS hearing which were meaningful. Yet Ford's undergraduate degree in economics disqualifies him as an "expert", while Kendall's Ph.D. makes him acceptable despite the rather wide gulf between high energy nuclear physics and the engineering of emergency core cooling systems. (It might be noted that even a Ph.D. in nuclear engineering is no automatic assurance that the holder has any detailed knowledge of emergency core cooling systems. One of the authors of this report is well acquainted with first-rate nuclear engineers whose professional interests and qualifications do not include the technology of core cooling systems.) The irony inherent in the arbitrary definition of an "expert" in the ECCS case did not escape the Friends of the Earth, who commented:

> Mr. Ford . . . [has] a new degree in economics rather than nuclear engineering--that makes him a non-expert despite

> his year of intensive study of nuclear safety. Another Harvard economist, ironically enough, is chairman of the AEC [James Schlesinger].[150]

The test of expertness applied in adjudicatory hearings (and in hybrid hearings such as the ECCS hearing) causes considerable procedural wrangling in the proceedings before AEC Boards. This puzzles some members of environmental groups:

> [In the Ford argument] perhaps the hearing board's fence-straddling can be accounted for by a strong disinclination to open the floodgates to "unqualified" witnesses, on the one hand, and on the other, a strong disinclination to be charged with stifling dissent. The position of counsel for the nuclear industry may be harder to fathom. They're playing to win, and if they really thought that Mr. Ford's testimony would be an embarrassment to their opposition, they'd be only too happy to have him testify.[151]

Now the reason for the ECCS hearing as well as individual licensing hearings is to develop a record upon which an adequate rule may be promulgated or a rational decision made. Would it not be better to allow the evidence of "non-experts" into the record (particularly in light of the difficulty involved in determining just who is an expert) and then to allow the ASLB (or the AEC) to give whatever weight it deems appropriate to such evidence. A basic tenet of the law of evidence is that every attempt should be made to avoid unnecessary confusion of a jury. But in this case we are dealing not with a jury but rather with a panel of adjudicators who are, in theory, quite capable of weighing the evidence presented. One might argue that the admission of "non-expert" testimony would increase the size of an already voluminous record. However, it would

----

150 ECO, Number 1, August 7, 1972, p. 8.

151 Ibid., p. 1.

seem that this could be avoided through judicious application of the principle enuniciated in Appendix A, Part V(d)(5) of CFR Part 2 which reads:

> (5) To prevent unnecessary delays and an unnecessarily large record, the Board may, pursuant to §2.757, limit cumulative testimony, strike argumentative, repetitious, cumulative, or irrelevant evidence. . . .[152]

(Any argument about the volume of the record, at least insofar as ECCS is concerned, is rendered moot by the fact that the transcript is already some 22,000 pages long. What then would constitute "an unnecessarily large record?")

One of the purposes of the AEC licensing procedure is clearly to arrive at a determination and resolution of the technical issues involved in the design, construction, and operation of nuclear power plants. This is not to suggest that all--or even the most important--questions involved in power plant licensing are directly related to the scientific and technological details of the plant. Obviously, the concerns of all involved parties are not merely with the technical efficacy of a particular technological application, but with the consequences of that application for individuals and society. It is equally clear, however, that determination of theimpact of a technology upon human and societal values can, and should, be based upon as definitive and accurate a description of the impacts of the technology upon its physical, natural, economic and social surroundings. (It must be emphasized, for it is a point frequently overlooked or misinterpreted, that the description and evaluation of impacts ought not to be confused with the act of arriving at a decision concerning the control or management of technology. The impact description and evaluation is merely one of many inputs to a decision which must ultimately be made through the weighing of conflicting interests and points of view.) In the current system of AEC proceedings questions of technical efficacy, safety, environmental impact and the like are often resolved (or at least addressed) in an adversary format--the same format, in fact, which is used to weigh and consider the consequences of power plants on individual and societal values. Is it reasonable to expect that something approaching a resolution of scientific and technological issues can result from such a format? It is interesting to note that, despite almost universal dissatisfaction with the current licensing and rule making procedure, there are those that argue for the maintenance, or upgrading, of a process of what has been called "adversary science." Thus, Mrs. William Carl of the Lloyd Harbor Study Group has argued that

> Because of the highly technical nature of the testimony in these hearings, the recent movement toward out of court exchange of written questions and answers should be stopped. Classic court procedures, with free cross examination and disclosure of witness demeanor evidence [sic] must be preserved.[152]

Irving Like emphasizes the need for cross-examination and other trial practices:

> The adversary posture is essential to achieve a dialogue between the citizen, the applicant, and the regulatory authority, the outcome of which is a body of knowledge tested in the crucible of cross-examination.[153]

There are those who propose the institutionalization of "adversary science." Donald Geesaman, a physicist at the Lawrence Livermore Laboratory of the University of California, has asserted the need for scientists with the "responsibility to prepare public adversary, or possibly advocacy, arguments"[154] to redress the imbalance created by the large resources of the proponents of

---

152 LH, p. 147.

153 Irving Like, "No-Win Strategy", at LH, p. 1408.

154 Donald Geesaman, "Adversary Science and Energy Decisions", Delivered at AAAS meeting, December 29, 1971.

technologies. In a recent article in the Bulletin of the Atomic Scientists, Geesaman and Dean Abrahamson, Director of the Center for Studies of the Physical Environment at the University of Minnesota, elaborate on the adversary science notion by proposing the formation of "forensic science groups" which

> would have the explicit and abstract responsibility to prepare critical arguments on some technologies which, in their judgment, would substantively affect the future of society.[155]

They argue that since in most cases there is an adequate articulation of the benefits of a technology presented by those who stand to gain economically from its application, it is necessary to provide a countervailing scientific position, through the "creation of a competent and significant body of contrary bias which would avoid the dangers inherent in assessment agencies of conjectured inviolable objectivity."[156]

There are, however, many who feel that the adversary system does not or cannot be an effective means for resolving technological issues. Attorney Harold Green, a believer in the adversary process for the resolution of issues involving social values, has commented that "there is little justification for a public hearing or . . . determinations as to the adequacy of technical details of a specific nuclear power plant."[157] Even a report of an *ad hoc* lawyers' group to the Atomic Industrial Forum recognized the uselessness of trial-type hearings for complex scientific and technical issues. That report stated that:

---

155 Donald Geesaman and Dean Abrahamson, "Forensic Science - A Proposal" in Bulletin of the Atomic Scientist, March, 1973, p. 17.

156 *Ibid*., p. 18.

157 LH, p. 341.

> . . . safety issues are . . . highly technical and traditional rules of evidence and of trial procedure become totally incongruous when applied to such issues as the adequacy of engineered safeguards systems. The rational approach here is much more akin to the techniques followed by scientific or engineering review panels than to the mechanics of trial procedure.[158]

Arthur Murphy points to the experience in AEC hearings as a justification for a change in the manner in which technical details are handled. In his Columbia Law Review article he comments:

> Where issues involve scientific data, written evidence rather than oral testimony should be preferred. No one who has observed an AEC licensing proceeding can doubt that examining a complex technical subject by means of oral testimony is both difficult and time consuming.[159]

Louis Mayo has described the difference between the scientific method and the adversary system as follows:

> . . . [S]cientific method is aimed primarily towards enlightenment, i.e., the production of knowledge, while an adversarial system is directed primarily toward power, i.e., the assertion of claims and the influencing of decision outcomes.[160]

In the Midland case, the attorney for Dow Chemical Company was explicit about the differing roles of science and advocacy. He noted that

> . . . the attorney is by definition and intention an <u>advocate</u> under our adversary system of law. He is paid to speak <u>for</u> his client, within the bounds of ethics and law. That is not, we submit, the role of the witness.[161]

---

[158] Atomic Industrial Forum, "A New Approach to Reactor Licensing", A Report by the Ad Hoc Lawyers Group; Robert Lowenstein, Chairman, in <u>LH</u>, p. 1447.

[159] Arthur Murphy, "The National Environmental Policy Act . . .", p. 996.

[160] Louis H. Mayo, "Scientific Method, Adversarial System and Technology Assessment", Program of Policy Studies, George Washington University, Monograph No. 5 (1970), p. 20.

[161] <u>Dow Memorandum Regarding Further Cross Examination of Bruce F. Watson</u>, July 15, 1971.

It is the difference in the goals of scientific method and the adversary system that leads us to the conclusion that for the purposes of determining technical efficacy and identifying the impacts of nuclear power plants on safety and on the environment, adversary science and the resolution of scientific issues in a trial type proceeding must fail to provide an adequate resolution of scientific and technical issues. The current process frustrates the resolution of technological issues, while providing a framework for legal and procedural gymnastics essentially unrelated to those issues. Adversary science, both as defined by its advocates and as practiced in AEC licensing proceedings, appears to reduce to the use of that scientific and technical information which supports a particular case and the discarding of any other information. While there is no question that at present the adversaries are not evenly matched, we do not conclude that the solution to the imbalance is to beef up the opponents of nuclear power and permit the resolution of technical issues in adversary proceedings. There is absolutely no evidence that, even in a balanced situation in which adversaries have equal resources and equal talent, the exposition of two (or more) positions arrived at by selective use of available evidence will lead to anything approaching scientific truth.

All of this suggests the need for independent study of the scientific and technical issues involved in nuclear power cases. Such independent study would serve as an input to a decision process which weighs the beneficial and harmful impacts of nuclear power plants (or a given nuclear power plant) and which almost certainly must include aspects of the adversary process. There is at present no institutional mechanism for such independent study.[162] As Ralph

---

[162] We will, in the section of this report on recommendations, propose a mechanism for independent study of the technical issues involved in the nuclear power controversy.

Lapp has commented:

> Too often, the intervenor has felt that he was in a contest with not only the utility and the nuclear vendors, but also with the Atomic Energy Commission. If local intervention is to serve as a check on deployment of unsafe reactors or on unsafe siting, then intervenors must have access to some independent authority with which to challenge the organized technical resources of the utility.[163]

If scientific and technical issues relating to nuclear power plants could be resolved (or at least discussed and described) by "technology assessment" teams of independent scientists (i.e., those with no direct state in the outcome of an assessment) then the forum for public participation would have as a basis an impartial evaluation of the impacts of nuclear power plants which could then be weighed by means of an adversary process. The fiction of attempting to resolve scientific issues through courtroom techniques and lawyers' strategies would be removed. Questions of the motives of scientists and the deliberate concealment of evidence would be eliminated. The entire proceeding would then approach the dual goals of arriving at scientific and technical determinations and allowing meaningful public participation in ways more likely to succeed than the present method of allowing an adjudicatory proceeding to do both simultaneously.

OTHER PARTIES TO THE HEARINGS

As mentioned earlier, intervenors depend to a large extent upon cross-examination and upon information supplied by the AEC and other parties to the proceeding for the scientific and technological inputs to their case since they are usually unable to do original technical work on their own. Thus, the

----

163 Licensing Amendment Hearings, p. 411.

strength of an intervenors case may well depend upon the adequacy and accuracy of the information provided by the applicant for a permit or license (or in the case of the ECCS hearings, by the vendors). But, as discussed in the last section, that information is selectively released in an adversary situation in which the utility or vendor and, by then, the AEC Staff, have a definite position which they are seeking to have adopted. It is logical to conclude then, that they tend to make such information available that will substantiate their case and withhold that which might negatively affect that case. Before applying for a license, the utility has made (at least to its stockholders and its management and directors) a prima facie case that the proposed plant is needed and that the plant can be built and operated safely and profitably. The vendors feel and assert that the equipment which they manufacture and sell is safe and effective. No one would seriously accuse manufacturers or utilities of deliberately planning to construct and operate unsafe plants or plants which entail too great a risk or unnecessarily degrade the environment. What is at issue in contested proceedings is whether or not society is willing to accept the inherent risks and the detrimental impacts on the environment; the difficulty of answering the question of what constitutes "too great" a risk or environmental impact is the reason that public involvement in the decision process is essential.

We have noted that utilities and vendors have available to them the financial and manpower resources necessary to make a comprehensive case and can utilize the services of large and sometimes talented groups of witnesses. That ability is important both to the proponents of nuclear power (since it helps them make their case in a strong and convincing manner) and to the intervenor opponents (since it provides them with a sometimes valuable source of information). But the fact that utilities and vendors are a primary source of information gives rise to very serious problems. There is a general reluctance on the part of vendors to provide information; quite often the manifestation of this reluctance is the claim that documents are proprietary and therefore, are protected under exclusionary provisions of the Rules of Practice and recognized by the Freedom of Information Act. Intervenors have contended that information with a direct bearing upon the health and safety of the public should not be permitted to be concealed by claims of proprietary interest. Some have contended that often the purpose of such a claim is not to protect trade secrets, but rather to serve as a psychological ploy to preserve a market position by giving the appearance of withholding important information which is supposed to imply that they have secrets worth preserving which, in turn, implies product superiority.[164] Myron Cherry attributes reluctance to subject all relevant information to public scrutiny to:

> . . . the many narrow interests of an industry which has placed its ability to develop and build hardware far ahead of its ability adequately to understand the significance of the problems related to the machines it builds.[165]

Another problem directly related to the role of utilities and vendors as information sources for intervenors is the potential for distortion and misrepresentation of information inherent in such a situation. Representative John Dingell (D-Mich.) Chairman of the House Merchant Marine and Fisheries', Subcommittee on Fisheries and Wildlife Conservation, has stated bluntly that,

---

164 Interview with Myron Cherry, May 15, 1972.

165 Licensing Amendment Hearings, p. 215.

"We can't outshout the utility lobby; it is one of the best financed in the country and capable of unlimited distortion."[166]

The adversarial nature of licensing proceedings almost requires that supporting scientific information be presented and opposing information ignored by each side in the controversy. And if intervenors are forced to perform analysis based on partial, false, or misleading information, the intervenors' analysis will undoubtedly suffer qualitatively and be of questionable validity. Nowhere was this point more clearly demonstrated than toward the end of a long day at the ECCS hearings when Henry Kendall, testifying for the Consolidated National Intervenors and the Union of Concerned Scientists, became aware that certain information (concerning the identification of fuel rods in photographs provided by General Electric) was misleading (perhaps deliberately). Some of Dr. Kendall's comments highlight the difficulties and frustrations which intervenor scientists and engineers must overcome when forced to rely upon such information to make and present their case:

> In the field of scientific research in which I have my professional career, the object of a professional paper reporting technical results is to report those results in a way that will make them accessible, free of ambiguity insofar as this is possible, to the other members of the technical community for their review and for their study, so as to solicit honest, capable and competent review of the material that is contained there.[167]
>
> . . . .
>
> . . . [T]he documents we have had to review in connection with the General Electric studies and researches into reactor safety have not been drawn up according to the standards which I consider appropriate or desirable or necessary for

---

166 National Journal, June 17, 1972, p. 1028.

167 ECCS Transcript, p. 18050.

> an independent, qualified, thorough review by other members of the technically trained community.[168]
>
> . . . .
>
> I find in the particular instance which has come to my attention that General Electric has failed to provide information which would clear up at first hand an ambiguity which could be expected to trap a person whi is accustomed to the tradition of full, honest, complete reporting of technical and scientific information.[169]
>
> . . . .
>
> I would say . . . there is an absolute obligation in cases where confusion can so easily arise . . . to leave each of these photogrphas clearly and definitively labeled so as to reduce the possibility . . . of confusion. . . .[170]

Attorney Myron Cherry, commenting on this episode noted that "no party has a right to intentionally obfuscate material."[171] General Electric Company responded to the charges by noting that

> The astounding acknowledgment, in effect, by CNI that it looked closely enough at the two information sources to observe a discrepancy no earlier than August 11 raises serious questions about the thoroughness of their review. . . .[172]

Nowhere in the General Electric comments is there any indication that GE sought, on its own volition, to eliminate the confusion caused by the documented testimony in question. The incident, we feel, raises serious questions of a violation of a public trust and a violation of the public's right to know. It also raises doubts about a major industry's candor while participating in proceedings

---

168 _Ibid._, p. 18051.

169 _Ibid._

170 _Ibid._, p. 18095.

171 _Ibid._, p. 18061.

172 _Opposition by General Electric Company to CNI Motion for Discovery and Taking of Deposition_, August 18, 1972, pp. 2-3.

of this sort. But, as we pointed out earlier, truth is often a casualty in a system better designed to present potent and convincing cases than to permit complete exposition of the facts.

Of course, not only scientific and technological experts are involved in AEC proceedings; if the scientists are misled by such tactics, what chance do lay citizens have? It appears evident to us that utilities and vendors have an obligation to make scientific and technical aspects of nuclear power understandable to the public. Harold Green has recommended that the applicant, as part of his application

> . . . include a full and candid disclosure, giving equal weight to optimism and pessimism, of all risks and benefits of the proposed plant in a manner and using language which will be readily comprehensible to the ordinary educated member of the public.[173]

But Arvin Upton, an attorney who has represented utilities, noted in a hearing that

> I despair frankly of ever being able to put these problems in terms which the average educated individual can understand. . . . I really think we fool ourselves in this country that all of these tremendously complex technological problems . . . will ever be reduced to terms where the average educated member of the public is either going to be able to understand . . . or is going to want to take the time.[174]

We disagree with this contention, although we acknowledge the difficulty inherent in explaining complex technical matters to laymen. The aim here, however, is not to make the public conversant with the details of nuclear power plant design, but rather to explain the issues in a manner which makes it possible

----

173 LH, p. 343.

174 LH, p. 355.

for the public, and for non-scientist decisionmakers, to evaluate the import of scientific and technical information and to use the information as an input to the decisionmaking process. (It is of some interest that the Courts have begun to take an interest in the need for understandable explanations of technological issues in environmental matters. In discussing impact statement preparation under the National Environmental Policy Act, a lower court has stated that the statement must be

> . . . written in language that is understandable to non-technical minds and yet contain enough scientific reasoning to alert specialists to particular problems within the field of their expertise.[175]

District Judge Bue of the U.S. District Court, Southern District of Texas, Houston Division notes, in another case that

> The reason for this standard is that impact statements must assist in rational, thoroughly informed decision making by officials higher up in the agency chain-of-command, including the Congress, the Executive and the general public, some of whom may not possess the technical expertise of those who evaluate the impact and prepare environmental statements.)[176]

However, for reasons explained at some length earlier, it is dangerous to make the simplified exposition of scientific and technical issues the responsibility of the applicant alone. Clearly, the straightforward and understandable presentation of those issues is a responsibility which all parties in the proceedings must share, a responsibility that, with few exceptions, none have accepted.

----

175 Environmental Defense Fund v. Corps of Engineers of U.S. Army, 348 F. Supp. 916, 933, (1972).

176 Sierra Club v. Froehlke 5 ERC at 1068 (1973).

Some brief mention is due, in this section, of the role played by the states in nuclear power plant licensing proceedings. The Director of Regulation of the AEC is required to send a copy of each license application "to the Governor or other appropriate official of the State in which the facility is to be located."[177] Pre-empted by the Federal government in the setting of standards for radiation emissions from plants,[178] the states have played their major role in the setting and enforcement of water quality standards. For the most part the states involved in the cases which we have observed have played a limited role in the exposition of scientific and technical issues (with the exception of water quality issues). Most of the scientific back-up for the attorneys representing the states comes from within the various state agencies, most notably fish and game commissions.

ROLE OF AEC

The Atomic Energy Act of 1954 (Public Law 83-703) establishes an Atomic Energy Commission to carry out the expressed policy of the United States that

> (1) a. the development,use, and control of atomic energy shall be directed so as to make the maximum contribution to the general welfare, subject at all times to the paramount objective of making the maximum contribution to the common defense and security; and
>
> b. the development, use, and control of atomic energy shall be directed so as to promote world peace, improve the general welfare, increase the standard of living, and strengthen free competition in private enterprise.

Since the passage of the Act, the AEC has been faced with the dual responsibility of promoting the development of nuclear power and of regulating that

---

177 10 CFR §2.101.

178 Minnesota v. Northern States Power Co., 405 US 1035 (1972).

development to insure public health and safety. Much has been said in recent years about the difficulty of both promoting and regulating a technology. The organizational charts of the Agency make clear that the AEC has attempted, at least functionally, to separate the roles. But the argument is often made that the difficulty is not one which can be alleviated by careful organizational arrangement but rather is one which is inherent in any agency with two such disparate goals. Roger Cramton, while chairman of the Administrative Conference of the United States, once made such a general argument:

> We all know that individual agencies tend to develop what might be referred to as belief patterns. They are concerned with particular missions. Their employees, often of a common background and with a common training, over time tend to see the agency's mission and the world through the same set of glasses, the same set of shared attitudes.[179]

Even the careful organizational arrangement cannot, or at least does not, avoid such problems as the placing of responsibility for safety research within the developmental part of the AEC. Representative Lester L. Wolff, in commenting on the nuclear safety research effort of the Atomic Energy Commission, noted that:

> I am . . . only too well aware of the inner drives of well-established bureaucracies, and remain to be convinced that the accumulated promotional momentum of the AEC can be so easily diverted from its previous course. For me, there remains a serious, discrediting conflict of interest in keeping the AEC's nuclear safety research and development under the direction and control of the same man whose feet will be held to the fire by nuclear power advocates if he does not quickly produce a breeder reactor demonstration for them. None of us, including Mr. Shaw [Director of AEC's Division

---

179 National Environmental Policy Act, Hearings before Senate Committee on Public Works and Senate Committee on Interior and Insular Affairs, March, 1972, p. 391.

> of Reactor Development and Technology], are supermen, and we cannot expect an Olympian sense of detachment from him as his right hand plans nuclear power research while his left hand plans the research that could raise new questions, even barriers to further acceleration of commercial nuclear power.[180]

Sooner or later, critics claim, in any matter before the Commission a choice must be made between the promotional and regulatory goals. Most critics claim, and we find the claim validated by our observation of the cases under consideration here as well as by interviews with Commission members and staff, that the choice is most often made in favor of the promotional goals. The attitude of some AEC Commissioners on the question of promotion vs. regulation is made clear in the following comments by James Ramey:

> . . . I'd like to emphasize that nuclear power will fulfill its promise. It is needed, and, in environmental as well as economic terms, it is the best hope for the world's power needs. It would be tragic if, because of a failure of public understanding, this promise were to be frustrated.[181]

Former Chairman Glenn Seaborg, testifying before the Joint Committee on Atomic Energy, expressed confidence that

> . . . people will come to understand that nuclear power probably presents the best method of generating the additional electricity that is going to be required in the future. . . .[182]

Similar quotations may be found in the text of many speeches by AEC Commissioners in the past several years. In its 1962 report to the President, the AEC spelled out its primary objective--an objective which has guided its actions to date:

---

180 Licensing Amendment Hearings, p. 129.

181 Speech to International Atomic Energy Agency, August 13, 1970.

182 LH, p. 21.

> The overall objective of the Commission's nuclear power program should be to foster and support the growing use of nuclear energy and, importantly, to guide the program in such directions as to make possible the exploitation of the vast energy resources latent in the fertile materials, uranium-238 and thorium.
>
> More specific objectives may be summarized as follows:
>
> 1. The demonstration of economic nuclear power by assuring the construction of plants incorporating the presently most competitive reactor types;
> 2. The early establishment of a self-sufficient and growing nuclear power industry; . . .
> 3. The development of improved converter, and later breeder reactors; . . .
> 4. The maintenance of U.S. technological leadership in the world by means of a vigorous domestic nuclear power program and appropriate cooperation with, and assistance to, our friends abroad.[183]

It is certainly true that intervenors and their attorneys see the Atomic Energy Commission as demonstrating a promotional bias in its handling of licensing and rule making proceedings. To some the conflict between promotion and regulation disqualifies the AEC as an objective decisionmaker in these hearings. Mrs. William Carl of the Lloyd Harbor Study Group claims that

> It is doubtful that any substantial improvement can be made in AEC regulatory procedures so long as the Atomic Energy Commission combines a promotional and regulatory responsibility.[184]

Anthony Roisman uses the argument of AEC's promotional bias to make the argument for meaningful public participation in the licensing process:

---

183 Civilian Nuclear Power--A Report to the President--1962, U.S. Atomic Energy Commission quoted in Civilian Nuclear Power--The 1967 Supplement to the 1962 Report to the President, U.S. Atomic Energy Commission, February, 1967, pp. 1-2.

184 LH, p. 149

> The AEC's haste in completing its review of nuclear power plants comes from an almost fanatical preoccupation with what the AEC sees as its role in meeting the so-called national electric power crisis. I desire to see the problems of a growing concern with environmental protection and public health and safety resolved. That resolution will not come where the federal agency charged with balancing those interests becomes completely captivated by one viewpoint. Certainly, so long as this one-sided attitude prevails within the AEC, the Joint Committee on Atomic Energy should do everything within its power to ensure full and unfettered _public_ participation in the AEC review process as a necessary check on agency action.[185]

Irving Like portrays the difficulties of a citizen intervenor in the face of the AEC's promotional bias:

> Since the AEC is the classic example of the military-industrial research and development complex, the limited resource litigator enters upon a David-Goliath type of confrontation pitted against the utility, the AEC technical staff and the titans of American industry. Arrayed on their side are also the huge complex of dependent trade association, economic interest groups, public relations media, the scientific, engineering and technical resources of the AEC, its national laboratories and sponsored research and the AEC's Congressional protectors.[186]

The dual role of the AEC is the subject of a suit filed in the District Court for the District of Columbia on January 6, 1972. Six citizen groups, the Conservation Society of Southern Vermont, the Chesapeake Bay Foundation, the Lloyd Harbor Study Group, Businessmen for the Public Interest, the Colorado Student Lobby, and the Courtlandt Conservation Association challenged, in the suit, the constitutionality of the Atomic Energy Act and its implementation. They claim that the promotional and regulatory functions of the Commission are contradictory and that the Commission's bias in favor of its promotional functions

---

185 _LH_, p. 465.

186 Like, "No-Win Strategy", in _LH_, p. 1402.

renders it incapable of being impartial and fair in its regulatory role. The groups ask the court to enjoin the AEC from continuing to exercise both functions. On March 3, 1972, the government filed a motion to dismiss and on September 6, 1972, the plaintiffs filed a response. The case is presently pending trial.

Rumors abound that a dramatic restructuring of the AEC is imminent. Most of these so far unsubstantiated speculations seem to agree that the regulatory function of the Agency will be spun-off from the promotional part. Some believe the latter will become part of a new Cabinet level Energy Department where it will compete with other forms of energy for the expenditure of agency research and development. We are not in a position, at this time, to evaluate the accuracy of these reports, but we do feel that the splitting of these functions cannot help but improve the opportunity for more meaningful citizen participation in the process and a more judicious consideration of the issues they raise.

By the time a permit or license application, or a proposed rule, reaches the stage at which the public becomes involved, the AEC's Regulatory Staff, which is charged solely with the regulatory, and not the promotional aspects of nuclear power, has already determined its position. Through what is claimed to be a careful review process, the Staff reaches a conclusion that the permit or license should be granted or that the proposed rule should be adopted. Thus, at the hearing, the AEC Regulatory Staff is in effect the ally of the applicant. Our observations bear this out; on virtually any question of import during all of the hearings we have seen, the Staff's position was the same as that of the Applicant. In fact, we have overheard AEC personnel involved in licensing cases, in private situations, admit that, "Of course, we're on the side of the applicant."

The established position of the Regulatory Staff is evident to all of the parties involved (although, understandably, it is of most concern to the citizen intervenors). Commissioner Ramey, in the course of an interview, indicated that he felt that the AEC staff should take part in licensing proceedings as an essentially neutral party, but he admitted that by the time a licensing permit goes to hearing, the Staff has already accepted the license application and ironed out most of the problems.[187] Martin Hoffman, formerly AEC General Counsel, also feels that the Staff "should get out of bed with the utility."[188] As the system presently operates, no hearing is held until the Staff agrees that the application meets AEC regulations. Thus, there is no representation of the broader public interest during the hearing by the Commission. To the extent that the AEC appears to represent the public, it is that part of the public that supports its mission and agrees with it or abstains from any comment on how that mission is implemented. If any other AEC representation of the public occurs, it does so in the private, off-the-record dialogue and review of the application which precedes the public hearings and which takes place out of the public view. George Freeman, who represents a consolidated group of utilities at the ECCS hearing, noted the solidification of Regulatory Staff position at the ECCS rule making hearings:

> All of us know, as a matter of common sense, that there are bound to be strong differences of opinion within any organization on highly complex and technical matters of great moment.

---

[187] Interview with James Ramey, November 6, 1972.

[188] Interview with Martin Hoffman, October 3, 1972.

> All of us with any connection with the nuclear industry have long known that there were a variety of views on the ECCS Interim Criteria within the AEC regulatory staff, within Oak Ridge, and within Aerojet. And yet, when the AEC Staff's prepared testimony was filed, we were given only the pros and not the cons. Accordingly, as the internal dissents within these organizations have been brought to light through cross-examination, they have attracted far more attention than they would have if they had been placed in an overall balanced context initially.[189]

The Staff's role in permit and license hearings is, in the view of Harold Green, one of the primary causes of a lack of public trust and confidence in the AEC. He notes:

> By the time a case is set for public hearing . . . the AEC Staff [has] already concluded that the construction permit or license should issue. The hearing itself is in reality staged for the purpose of obtaining a ratification of that decision. Perhaps the single greatest reason for the credibility gap is that in public hearings the AEC itself is always seen by the public as vigorously defending the applicant and fighting the intervenor, playing the role, as many people perceive it, of an enemy rather than a protector of the public interest.[190]

In summary, substantive due process cannot be made a meaningful actuality where intervention takes place in a tribunal representing an agency whose staff has already come to agreement with an applicant, where the applicant has already adopted the agency's recommendation, where the legislative mechanism (Congress) has already committed, without significant and open debate, the resources for promoting the technology and where the basic policy is not subjected to open, full, free, and fair debate in a broadly representative public forum before a citizenry informed of the issues.

---

189 George Freeman, "The Emerging Role of Rule-Making in the AEC Decisional Process", April 17, 1972, p. 8.

190 LH, p. 342.

The environmental impact statements filed by the AEC under the requirements of the National Environmental Policy Act provide a potentially valuable source of information for all parties involved in AEC proceedings. NEPA and its interpretations have been the subject of a large number of analyses and court decisions. It is not our intention to repeat the arguments made in these analyses. But we do feel it necessary to at least briefly discuss the role which NEPA has played in bringing information before interested citizen groups. Our conclusion, it will be seen, is that the environmental impact statements have been less valuable than they might have been in providing objective and meaningful inputs to citizens. (This conclusion should not be taken as a criticism of NEPA or its purposes. On the contrary, we feel that it has provided a vehicle for a national expression of concern over the wanton and haphazard disregard for environmental resources that has characterized our technological and economic development.) In our opinion, the failure of impact statements as information sources for citizens can be traced to several causes: the bias inherent in their preparation, the overwhelming concern with procedural as opposed to substantive characteristics of the statements, and the emphasis on the physical as opposed to the social and human environment.

The NEPA-required environmental impact statement is prepared for the AEC Regulatory Staff, usually by groups of scientists at National Laboratories. We have already discussed the Staff's role in licensing proceedings; at some point the Staff becomes, for all practical purposes, a proponent of the nuclear plant in question. For this reason, the Staff is quite likely to fail to recognize the importance of serious environmental problems. Further, it is clear from the AEC's <u>Guide to the Preparation of Environmental Reports for</u>

Nuclear Power Plants that the "Applicant's Environmental Report", which is discussed therein, forms a major input to the Staff's impact statement. Frederick Anderson, editor-in-chief of the Environmental Law Report, has characterized past AEC impact statement preparation procedure as follows:

> Under AEC rules the applicant initiates the review of environmental factors. His lack of enthusiasm for identifying and researching potential environmental problems in connection with his license is likely to be considerable, yet among the relevant parties no other is accorded more time for preparing its statement, or more influence over the content of the AEC draft statement. The AEC does not require its officials to make an independent assessment of environmental effects. Sister agencies have one month--20 working days--in which to comment; state and local agencies have twice as long. Environmental effects are not considered at hearings unless raised by intervenors, or applicants. When issues are raised, the AEC board will make findings. At no point does the AEC specifically provide for an independent, agency-originated assessment of environmental impact, as contemplated by NEPA.[191]

It comes as no great surprise that the applicant's report invariably finds that the particular site and plant is preferrable to all alternatives. No applicant's report or NEPA impact statement recognizes negative environmental impacts serious enough to recommend that a plant not be sited as planned. Thus, the Midland impact statement concludes that "the benefits to be derived from operation of the Midland Plant Units 1 and 2 outweigh the adverse effects identified in this statement."[192] The Vermont Yankee statement determines that:

> On the basis of the analysis and evaluation set forth in this statement, after weighing the environmental, economic, technical and other benefits of the Vermont Yankee Nuclear Power Station against the environmental costs and considering available

---

191 Environmental Law Report, February 1971, Quoted in Senate Committee on Public Works and Senate Committee on Interior and Insular Affairs Hearings on NEPA, March, 1972, p. 439.

192 Final Environmental Statement Related to Construction of Midland Plant Units 1 and 2, U.S. Atomic Energy Commission, Division of Radiological and Environmental Protection (March, 1972) p. iii.

> alternatives, it is concluded that the action called for is the issuance of an operating license for the facility.[193]

This finding was made subject to four conditions, three of which, in effect, require careful monitoring and study during operation and the fourth of which places a limit on residual chlorine concentration. Whether or not these conclusions are proper, given the circumstances surrounding the plants, what is notable about the conclusions of the environmental statements is that one could have predicted the substance of the conclusions, if not the precise language, even before the statements were filed. If nothing else, this creates a severe credibility problem, and greatly hampers the use of impact statements as information sources for citizen groups and as a basis for evaluating environmental impacts. If more effort were expended in assuring an unbiased, impartial, independent environmental statement, at least one barrier to effective utilization of the impact statement would be removed.

Two incidents occurred during the course of our study which we found curiously interesting, but which are difficult to analyze. They might, though, be of some value in understanding the views of some people involved in the NEPA process. On one occasion, a scientist, who led an environmental impact statement preparation team at a National Laboratory, spoke with us rather briefly and reluctantly about the impact statement and the genesis of data upon which conclusions were reached. When we asked to spend a longer period with him he responded that he would have to get clearance from "the director" although it was not at all clear to whom he was referring. His demurral made it clear that

---

193 Final Environmental Statement Related to the Operation of Vermont Yankee Nuclear Power Station, U.S. Atomic Energy Commission, Directorate of Licensing (July, 1972) pp. iii-iv.

he did not care to discuss the impact statement preparation process with us--or if he did, that he felt some compelling necessity to have clearance to do so. We handed him our card and asked if he would try to arrange such clearance and contact us as to when we might interview him. No subsequent contact was made. On another occasion, a government official in charge of the preparation of impact statements for his agency (not the AEC) assured us that where data were not available to complete an impact statement, the agency proceeded to invent the data for inclusion. Whether this is true or not we have no way of evaluating, though we saw no reason for the official to make the statement at the time unless it were indeed true. Whether or not the experience is generalizable to other agencies is not clear.

One sign that the AEC is attempting to make its impact statements less reflective of a promotional bias can be seen in the final impact statement for Indian Point Generating Plant Number 2. While concluding that benefits outweighed costs and recommending the issuance of an operating license, the Regulatory Staff also recommended that a "closed-cycle cooling system shall be designed, built and placed in operation no later than January 1, 1978"[194] in order to avoid huge fish kills at the plant intake structure. It is too early to tell whether attacking such a quid pro quo to the Indian Point Statement is an exception, the beginning of a trend, or an effort to diminish the considerable pressure being brought on the Agency, some members of Congress, New York State Officials and the much battered executives of Consolidated Edison Company (Con. Ed.).

----

[194] Final Environmental Statement, Indian Point No. 2, September, 1972, p. vii.

The predominant concern with procedural aspects of the impact statement arises from the fact that, at least until very recently, NEPA has been interpreted by the courts as an essentially procedural statute with no apparent substantive obligations. (As noted earlier, recent lower court decisions appear to be expanding the scope of the Act, but at present none of these decisions has had a demonstrable effect on AEC procedures.)

The third cause of the failure of impact statements as information sources is related to the emphasis in the statements on the physical environment. "Environment" is never explicitly defined in NEPA. It is clear that many Federal agencies (the AEC among them) have chosen to define "environment" in a narrow sense to include primarily the physical surrounding of a given project or facility. But the Act, and in particular, section 102(2)(C), which requires the preparation of impact statements, refers to "the quality of the human environment." This phrase can be broadly defined to include the physical, social, economic, political, and psychological ambience within which people exist. An impact statement examining the consequences of a given action upon such a broad "environment" becomes a "technology assessment", an assessment of overall consequences of decisions, programs, and projects. Such comprehensive "technology assessments" are lacking in the field of nuclear power (as they are in most other fields, as well). Considerable interest in the development of the concept of technology assessment and its application to the decision process underlying the commitment of national resources to a particular technology has been, for several years, debated in the Congress. Last fall (1972), Congress created an Office of Technology Assessment to provide itself with analysis of the potential impacts of technology. In that the OTA has yet to begin to function, it is too early to evaluate its impact upon the development of the technology assessment concept and upon Congressional decision-making.

We have not discussed in detail the role played by the Advisory Committee on Reactor Safeguards in this report. The only public representations of ACRS before March, 1973 were contained in a series of almost standard form "boiler-plate" letters to the AEC assuring the Commission that a particular nuclear plant was safe and that any problems with the design of the plant could be easily resolved (see Chapter III, Sections on Midland and Vermont Yankee). In March, 1973 the ACRS reluctantly opened parts of its meetings to the public after being forced to do so under threat of suit. At least until then, the ACRS provided no substantive information which could have been used by citizen groups; whether this will change in the future remains to be seen. It is notable that the ACRS, a committee comprised of scientists, would have felt the necessity of conducting their review of individual nuclear plants in secret. Intervenor groups have raised the question as to why they insisted on hiding their deliberations from the people. It seems a valid question to raise and interestingly enough, it is one that has never been given a satisfactory response.

CHAPTER V: SUMMARY OF FINDINGS AND RECOMMENDATIONS

Since, as a nation, we have become increasingly dependent upon technological development for our economic well being, it is inevitable that the impact of technology has become an increasingly evident force affecting the everyday life of our citizens. Few, if any, Americans are immune to the consequences of some aspect of technology, consequences which strike close to the core of their existence and to the quality of their living. As a result, public awareness that the benefits of technology are frequently accompanied by negative impacts has spread widely. And as that awareness has grown, it has been accompanied by increased citizen demands for a significant role in the decisionmaking processes which underlie policies and investments, particularly, but not only, in the public sector. Decisions about the applications of technology often involve considerations of human and societal values in which citizens have begun to seek a greater voice.

Our study, carried out over a 15 month period, observed and analyzed the marshalling and use of scientific and technological resources by some citizens in order to influence governmental decisions about nuclear electric power generation. Of course, this is only one area of citizen concern about the impact of technology on their lives. While some aspects of our study and its findings focus exclusively on actions brought before the Atomic Energy Commission, we feel that major parts of the study are relevant to a broader range of issues associated with the frequently unanticipated effects of technology and rapid technological change on the physical and human environment. We structured our research to learn how citizen groups, active in what has been called the "nuclear power controversy", obtain and utilize scientific and technological information in formulating and presenting their cases before a Federal administrative agency (the Atomic Energy Commission) which makes decisions on standards, criteria, and the granting of construction permits and operating licenses for nuclear power plants.

Citizen group involvement in the AEC's hearing process has normally been directed at one or more of three generic issues: nuclear safety, environmental impact, and agency procedures. We have made no judgments about the merits of the various contentions involving the first two of these issues but our detailed examination of the Commission's decisionmaking process allows us to comment on the third. In general, we have found that the requirements of procedural due process are scrupulously attended to by the Atomic Safety and Licensing Boards (ASLB's) which hold hearings on behalf of the Commission and by the Commission itself. On the other hand, we have found that, because of the composition, predilection, and defined role of ASLB's, citizen group opponents of nuclear power plants are denied substantive due process and, therefore, meaningful participation with the opportunity to affect decisions concerning the licensing of nuclear power plants. (The role of ASLB's is to hold hearings to ascertain whether an application for a permit or license meets AEC standards and National Environmental Policy Act requirements and not to determine whether or not a plant should be built, whether or not it is needed, and whether or not the community where it is located wants it). These determinations are made by the utility involved and are based on their own predictions of need and economic viability.

Citizen participation in the process we have witnessed takes place at administrative hearings at three levels: generic or rule making hearings (we have closely followed the emergency core cooling system and nuclear fuel cycle proceedings); construction permit hearings (we have examined, in detail, the proposed Midland, Michigan nuclear power plant hearings and, in somewhat less detail, several others); and operating license hearings (we have examined, in detail, the Vermont Yankee nuclear power plant hearings and, in less detail, several others).

We should observe, at this point, that the AEC has distinguished itself among Federal regulatory agencies in its efforts to provide opportunities for citizen participation in its hearing process. Nevertheless, to the extent that the AEC hearing process is designed to be an adequate and fair forum for the expeditious public consideration of an application to construct or operate a nuclear power plant, it is a failure; to the extent that it is designed to allay community fears (such as those represented by intervenor opponent groups), it is a failure; to the extent that it is designed to resolve scientific or technological questions or issues, it is a failure; to the extent that it is expected to resolve environmental issues, it is a failure; to the extent that it is expected to satisfy the legitimate needs of applicants, proponents, or opponents by demonstrating the validity or lack of validity of their contentions, it is a failure. Throughout the course of our study, we found few individuals who felt the process was anything but a failure. The positive comments we did hear argued that, even though they provide a poor forum for objective or independent consideration of issues, the AEC Rules of Practice at least lay out a clear legal path for opponents to follow in order to carry out their intervention. This may be true, but the only consensus among all of the parties to the proceedings appeared to be a general evaluation that the whole process as it now stands is nothing more than a charade, the outcome of which is for all intents and purposes predetermined.

When a citizen group files a petition for leave to intervene for the purpose of opposing the construction or operation of a nuclear power plant in its neighborhood, it has, in essence, decided to enter a struggle against huge odds. In fact, at operating license hearings, the odds are seemingly insurmountable, given the fact that a plant has already been built and a large capital investment already made. (It is of interest to note, however, that many of the more significant interventions have, until quite recently, occurred at the operating license stage.) Citizen groups, then, take part in interventions despite these odds, or because they are oblivious to them. The groups hope to achieve their defined goal: to stop the plant's construction or operation altogether, to delay the construction or operation in hopes of raising costs to the point at which the applicant utility company is willing to make significant concessions, or to assure that the plant strictly adheres to AEC standards and regulations and is demonstrably a safe facility incorporating the latest in safety and environmental control technology and that will cause minimum environmental disruption.

Some citizen groups claim that public education about the effects of nuclear power is an underlying purpose for their involvement. But we have been unable to learn from these groups or their leadership why they view public education as so important and why they allocate such a significant portion of their severely limited funds toward that end. One expected outcome of their public education programs seems to be to attract converts to their cause and thereby to obtain additional contributions to support their activities.

Our observations, substantiated by a number of other studies on public attitudes toward nuclear power, indicate that citizen groups are motivated in large measure by fear--primarily fear of accident and fear of radiation exposure.

For the most part, the opponent intervenor groups in individual nuclear plant cases are local voluntary associations aggregated for the purpose of intervening in the hearing process and perhaps, subsequently, for appealing to the courts the outcome of that process. In some individual licensing cases and in rulemaking hearings, national groups, often environmental organizations and in a few cases, labor unions, have joined the opposition as intervenors. In those cases, though, the national organizations are generally primarily interested in making a case on generic issues which relate to all nuclear plants. But usually, we have found the opposition intervenors, and thus the interventions, to be essentially local in nature, locally inspired, locally funded, and directed toward local concerns and issues. The groups we have observed claim to represent only themselves and do not claim to speak for the community at large. The existence of the groups depends directly upon their ability to raise funds through voluntary contributions and to maintain the support of their membership (or constituency).

Since the AEC's hearings provide a focus for citizen involvement and since active participation in the hearings requires the services of a lawyer, one of the first actions of a citizen intervenor group is the hiring of a lawyer. Lawyers representing citizen groups are often those who have represented other groups in similar nuclear power cases. They have the responsibility of presenting the case; they isolate the particular issues to be raised and structure the presentation according to their own strategies and tactics, which are strongly influenced, if not dictated by their own talents, perceptions, techniques, and indeed, shortcomings. The citizen groups do the legwork, copy documents, provide the organizational impetus and perform the public education and public relations functions.

In general, lawyers deal with and are responsible to the (elected) leadership of the citizen groups and almost never see or meet the whole membership.

Citizen groups often have their own scientific advisors, usually unpaid volunteers who raise issues and help to prepare certain written materials on scientific and technological issues. If the lawyer requests and funds are sufficient, the citizen groups also arrange for paid technical consultants to help him review the numerous relevant documents, write questions for cross-examination, and, in some cases at least, to interrogate witnesses on technological issues. But for various reasons already discussed, competent technical assistance is difficult, and often impossible for them to find. Thus, the volunteer scientists and engineers often have little or no training, experience, or expertise in areas directly relevant to nuclear physics or nuclear engineering. Paid consultants are frequently graduate students in nuclear engineering or in various environmental fields and usually have no experience in nuclear reactor design and operation. Expert witnesses appearing on behalf of citizen groups often possess a disciplinary training not directly applicable to the issues at hand. In fact, the areas of expertise of most scientists and engineers are likely to be sharply limited and discrete. But the issues involved in the licensing of nuclear power plants are broad and interdisciplinary in nature and thus beyond the expertise of any single individual. We have found little understanding among citizen groups that problems associated with nuclear power plants, as well as most environmental problems, are essentially interdisciplinary and thus require, for their understanding and resolution, interdisciplinary impact analyses or technology assessments. But even if citizen groups fully appreciated this concept, interdisciplinary assessments are by nature expensive and, to the extent that these groups depend upon voluntary contributions and are constantly short of funds, beyond their means.

On the other hand, we have found that the AEC, the applicant, and some proponent intervenors (notably primary beneficiaries of a proposed plant and equipment vendors) have available to them and utilize a full range of scientific talent in the physical sciences and engineering in preparing and presenting their testimony. It is apparent that these parties, with the exception of the AEC, have a huge economic stake in the outcome of the hearings. The licensing hearings are, after all, only one portion of an expensive process that is encountered by proponents of a nuclear power plant; construction of the plant requires the expenditure of hundreds of millions of dollars. The applicant typically spends several hundred thousand dollars to pay scientific and engineering analysts, witnesses, and consultants. These costs are ultimately passed on to consumers as part of the rate base or are integrated into the price of construction. These are, after all, legitimate costs of doing business and can ordinarily be easily absorbed.

The imbalance in scientific, technological, and economic resources between the proponents of nuclear power and their opponents make it much more likely that the proponents will make a convincing case in support of a proposed plant than that the opponents will present compelling arguments against it. At the same time, it is important to recognize that no one involved in the process makes judgments about the issues (even those issues which are almost purely technical) from an independent or disinterested perspective.

One major problem arises because citizens do not become involved in the nuclear power plant licensing process until quite late. Before they can take an active role, a preliminary application and safety analysis must be submitted by the applicant to the AEC Regulatory Staff; these documents are considered by the Staff which identifies shortcomings and data deficiencies; the applicant then discusses the application with the Staff and takes the necessary steps to provide the requisite information, redesign plans, perform additional analyses and studies, and resolve those issues raised by the Staff; a final application is then submitted; with the agreement of Staff, applicant and equipment vendor about the details and suitability of plans for construction (or operation) of the plant, a hearing schedule is announced by the Atomic Safety and Licensing Board. It is only at this point that opponent intervenors acquire access to the process. Applicant and vendor have been working on the process for years and the AEC staff has reviewed the plans extensively for months; but the intervenors begin an assessment, having seen little of the documentation and having almost no detailed knowledge of plans to that point. Their late entry and the difficulty of obtaining all of the relevant documents combines with other factors to prevent intervenors from presenting a fully supported and convincing case during the hearings.

In short, the process, in our view, is neither fair nor expeditious nor democratic in any substantive or meaningful way. It is marked by the manipulation of scientific and technological information by all parties in order to substantiate their predetermined points of view. It is a system evidently designed by lawyers for the consideration of legal issues and as such does not serve to resolve important issues of science and technology and of human and physical ecology which are likely to accompany the construction and operation of nuclear power plants. The process is characterized by the confrontation of a special interest group (utilities)--affluent, influential, deep in manpower and financial resources, supported in its contentions by statistics of growing population and increased consumption of electricity and by recollections of the brownouts and blackouts of recent years, and aligned with other powerful (vendors) interests and with a Federal agency (AEC)--and an ad hoc, underfunded, disparate group of citizens seeking to ward off or otherwise influence the imposition of a technology which they feel is unacceptable and inherently dangerous.

The AEC licensing process provides no effective mechanism for the independent analysis of the benefits or hazards of the construction and operation of any given nuclear power plant at any given site. But despite this major shortcoming in the existing process, citizen groups have raised a number of important questions which might never have been aired in a public forum had they not been willing to make the personal sacrifices their participation entailed.

The relative ease of access to the AEC hearings, while providing a means for them to express themselves, does little or nothing to diminish their feelings of powerlessness in the face of decisions which will impact on their lives. Their participation can be characterized as a manifestation of their anger and frustration at what they interpret as a hollow extension of procedural due process and the denial of meaningful or substantive due process. The result is that the hearings are, in the view of some citizens, expensive wars of attrition, designed to obfuscate rather than clarify the true risks and benefits of nuclear power.

Many members of the public have characterized the AEC's attitude as indicating that nuclear technology and decisions about its adoption, are too complicated for them to understand or worry about and that the agency can be relied upon to care for their health and welfare. Much evidence indicates, though, that a large number of governmental entities do not attend to the public's health and welfare, but rather are keyed to special economic interests and serve special constituencies, often to the detriment of the general welfare. Credibility has become a major issue and large and growing segments of the population remain skeptical, not only of the representativeness of the Atomic Energy Commission, but of governments on all levels.

It is not difficult to see citizen disaffection with nuclear power as a symbol of increasing dissatisfaction among some segments of the population with the economic and technological determinism that they feel has characterized governmental management of limited environmental resources and of a broader and more pervasive dissatisfaction with governance itself.

FINDINGS

Procedures

1. The Atomic Energy Commission asserts that its public hearings are designed to serve three purposes: to provide an orderly process for the licensing of nuclear power plants; to raise and review scientific and technological issues; and to provide a forum for public participation. Experience does not indicate that a single proceeding can adequately serve all of these purposes, nor would evidence suggest that any of them is well served by the present system. [p. 139]

2. During the early 1960's, public hearings were viewed as a process to win public support and to explain nuclear power and only incidentally to consider or resolve scientific or technological issues which the Agency felt were being adequately addressed by internal Regulatory Staff review of applications for construction permits and operating licenses and by periodic review of criteria. [p. 139]

3. All parties involved are critical of the procedures but from differing points of view: utilities are critical because the process does not adequately serve as a mechanism for considering salient scientific and technological issues but instead, often serves as a medium for delay which increases the lead time for getting a plant on-line and thus, the costs of construction; citizen group opponents are critical because they see the hearings as an elaborate attempt to give the appearance, but not the reality, of meaningful public participation. They also feel that the important issues related to nuclear power are not adequately considered by the hearing Boards. [pp. 140-141]

4. ASLB hearings are legal proceedings, evidently designed by lawyers for lawyers, and as such are ill-designed to deal with the problems of arriving at scientific or technological truth, to resolve issues of nuclear safety, or to provide a greater understanding of potential environmental impacts. [p. 142]

5. Though changes within the AEC in the last two years have tended toward the accommodation of citizen interventions, new policies, like the old ones, extend procedural due process while not changing the basic structure of the hearings to permit citizen group input to play an important or meaningful role in the decisionmaking process. These changes in form have not as yet brought about any discernible changes in the substance or quality of the interaction among parties to the proceedings. [p. 143]

6. Hearings represent only one of many causes for delay in the licensing of nuclear power plants and time spent in hearings represents only a small part of the total time elapsed between application and the granting of a permit or license. Moreover, in most cases, accusations that citizen opposition intervenors are directly responsible for increasing the length of hearings are not substantiated. [p. 145 ff.]

7. Protracted hearings place a financial burden on all parties. Thus, there is real danger that hearings will be reduced to a war of attrition--a test of staying power in which the party with the most adequate resources will inevitably prevail independent of the validity of its case. Obviously, citizen opposition groups, which rely on voluntary contributions by supporters who can realize no financial benefits from their donations, are at a distinct disadvantage when delays occur. [p. 151]

8. There has been widespread criticism that the holding of contested hearings at both construction permit and operating license phases causes undue delay and constitutes double jeopardy for the applicant (who, by the latter stage has already made a huge investment in construction of a plant). To date, constant changes in statutory and legal requirements, as well as in plant design, new technologies, operating experience, and deviations from quality assurance programs, have warranted the duplication of hearings. It would be preferable and expeditious to have a single set of hearings but this cannot be considered a reasonable means for considering applications as long as the courts and the Atomic Energy Commission itself continue to alter the ground rules under which licensing will take place. [pp. 151-152]

9. Changing procedural and substantive rules not only make a single hearing system difficult to envision, but also lead to delays in the current hearing process as well as to uncertainties about the obligations of each of the parties to the proceedings. [p. 152]

10. Contentions raised in individual cases often involve matters which might better be handled in generic proceedings. This would permit individual hearings to deal with particular problems specifically related to individual cases. It would also increase the difficulty of insuring adequate public participation because it could lead to the centralization of hearings in the Washington area, removing the issues from local attention and attendance. [pp. 153-154]

11. Presently, there is no public mechanism providing a realistic check on the applicant's fulfillment of commitments incurred during the construction permit hearings or at the licensing hearing stage. The same ASLB does not

usually hear both cases and follow-up responsibility is not vested in any single board. The Regulatory Staff, moreover, may not see a particular commitment as an important one and where the commitment is not made a specific condition of the issuance of the license or permit, the Staff may not feel constrained to insist on enforcement. [p. 154]

12. The question of burden of proof, although clearly dealt with in the AEC regulations, remains unresolved in practice. The regulations place that burden upon the applicant; in hearings, however, rulings of ASLB's about burden of proof often shift responsibility to intervenors or to the AEC Staff. [p. 154]

13. Much of the delay in the hearing process can be attributed to legal and procedural hassles concerning production of documents and responses to interrogatories. Much of it results from a need to prepare responses, briefs, motions, rulings, technical reports, and environmental reports, and to set suitable hearing dates to accommodate part-time Board members and all of the parties involved and to avoid conflict with other concurrent hearings. [p. 157]

14. Even if full disclosure became the rule, there is no evidence to suggest that citizen intervenors would be capable of understanding, interpreting, and evaluating the information made available to them, particularly where such documentation exists in great abundance. [p. 159]

15. The inability of intervenors to obtain documents, information, and responses to their interrogatories early in a given proceeding has, at times, caused great difficulty in their ability to develop their case. [p. 161]

Role of Lawyers

16. The administrative hearings of Atomic Safety and Licensing Boards provide a focus for citizen group intervention which establishes a pre-eminent role for the lawyer. For the most part, the citizen is reduced to being an observer of the proceedings--a non-participant who must depend upon others for a clear articulation of his interests. [p. 162]

17. Few, if any, lawyers involved in nuclear power cases have a real understanding of the technical details of nuclear technology by virtue of formal scientific or engineering training or experience, although some of them possess a clear view of the critical issues. [p. 166]

18. Lawyers meet the problem of lack of expertise about the technical details in different ways. Some attempt to handle interrogation of witnesses on technical issues without assistance; others make certain that they are accompanied by scientists or engineers to aid in their presentation of cases. Some lawyers see a limited role for scientific assistance and assign technical advisors a behind-the-scene or briefing role though others give technical assistants considerable leverage and even an active role as cross-examiners on technological issues. [pp. 168-169]

19. Some lawyers freely admit that they are severely hampered by lack of scientific and technical understanding or assistance though some attempt to obscure their lack of understanding of the scientific or technical details of nuclear power be becoming fluent in the jargon of the field. Lawyers attempting to carry out technical interrogation on their own are sometimes likely to ask unrelated, distorted or meaningless questions. Often, they demonstrate little or no understanding of whether or not an answer is valid, makes sense,

or opens other important avenues of inquiry that should be explored. It is not unusual to observe incorrect or unsubstantial responses to questions going unchallenged. [pp. 167-168]

20. Arrogant and often contentious behavior by lawyers (representing all parties) unnecessarily polarize the hearings. Meaningless questioning and wrangling by lawyers for all parties frequently obfuscate the real issues, delay the process, and accomplish little by way of resolution of outstanding questions. [p. 171]

21. Intervenors often use the hearings to test agency process rather than to examine technical fact. Non-substantive, procedural issues dominate the hearings to no discernible public advantage. This may result from the fact that, as noted in Number 17 above, lawyers lack understanding of technical details and are more comfortable dealing with procedural rather than with substantive scientific matters. [pp. 172-175]

22. Most of the lawyers, in the cases we have examined, freely discuss their role as "hired guns" representing only their client's interests and are not necessarily always fully convinced of the rectitude of the positions they represent. [p. 164]

23. Our own subjective evaluation leads us to the conclusion that, with some evident exceptions, the level of legal competence among lawyers for all parties is high and that their knowledge of the rules (and loopholes) is extensive. [pp. 162-177]

24. On occasion, technical interrogators have been used by some parties to make cross-examination more meaningful. Experience shows that sometimes technical interrogators do serve that purpose, while at other times they contribute

to delay and obfuscation of the issues. Frequently, technical interrogators find themselves questioning in areas where the lines separating the scientific from the legal are unclear. At other times, technical interrogators tend to outrun their areas of expertise and probe in areas of science or technology with which they may be no more familiar than lawyers who have little or no scientific training. [pp. 176-177]

ASLB

25. Lawyers serve as chairmen of the ASLB's and while their presence, along with two technical members, is to referee procedural matters, their rulings cannot always be directed exclusively at such issues. It seems inevitable that their judgments will impinge on technical matters, including testimony of witnesses and arguments of technical interrogators. Often chairmen of these Boards do not adequately understand the technical issues involved. [p. 42]

26. Atomic Safety and Licensing Boards approach licensing hearings with a predisposition to approve an operating license or issue a construction permit for the plant in question. Atomic Safety and Licensing Boards cannot be considered independent or objective adjudicative bodies, dissociated from the agency to which they are responsible and of which they are an integral part. It is to that agency that their Initial Decisions are transmitted for final decision by the Commissioners. [pp. 178-180]

27. The Atomic Energy Commission has recently been faced with formal charges of bias on the part of members of Atomic Safety and Licensing Boards, many of whom are full-time agency employees and others of whom are grantees or recipients of contracts from the AEC. The Commission has handled these

charges in a manner which does not eliminate the problem or in any way dispel the impression of predetermination of cases by the ASLB's. [pp. 180-187]

28. Atomic Safety and Licensing Boards have no planning function in the siting or construction of nuclear power plants. They only consider whether or not construction plans or operating license applications meet Atomic Energy Commission regulations and criteria. [pp. 6-7, 178]

29. ASLB's are called upon to judge the adequacy of the staff-prepared environmental impact statement mandated by NEPA [Section 102(2)(C)]. Until recently, no evidence could be found that the technical members of the Boards were particularly well prepared by virtue of their training to make such evaluative judgments which they had been making and in many cases still do. In the past, consideration of fulfillment of NEPA requirements has been directed toward compliance with procedures and not to any great extent toward the examination of substantive impacts. In recent months, however, the Commission has opened its panels to include members with training in some areas of the environmental sciences, particularly the biological sciences and environmental engineering. It is not yet possible for us to make a meaningful judgment about the Commission's new arrangements to deal with environmental issues as new procedures and regulations have only recently (in the past six months, or so) been issued. [pp. 177-178]

Role of Citizen Groups

30. Widening recognition in recent years of the pervasiveness of the impacts of technology has resulted in skepticism about the ability of "experts" to anticipate the full range of these impacts and of government and the private

sector to make decisions about the use of technology which fully reflect the public interest. As a result, citizens have sought a greater role in decisions about technology's use and about the diversion of limited resources to support its development. [p. 188]

31. At least part of the motivation for citizen involvement in nuclear power cases is a reaction to the dominance of experts and the apparent powerlessness of citizens to affect decisions about the expenditure of tax dollars to support technologies, the consequences of which are not fully understood. [pp. 189-190]

32. Citizen groups participate in nuclear power cases for reasons which tend to fall into three categories: a concern that the nuclear power plant be made safe (or as safe as possible) and that detrimental impacts of its operation be minimized; a concern that public information be upgraded; and an attempt to delay or stop the plant. [p. 190]

33. The concern of citizen groups has led the AEC to undertake a series of "experimental" hearings for the public airing of generic issues such as emergency core cooling system, radiation emission standards, and environmental effects of the nuclear fuel cycle. [p. 190]

34. Although many citizen groups profess a goal of improved public education, public hearings (the focus of citizen group interventions) are not well attended by the public and do not seem to significantly increase public understanding of nuclear power. A popular ground-swell to support citizen group efforts has not developed despite widespread public concern for the potential effect of nuclear accidents. [p. 192]

35. Of the citizen groups observed in this study, only one claimed that delay, and, in fact, cancellation of plans for the construction of a particular power plant were its goals. [p. 193]

36. Citizen intervention leads to some delay in the licensing process but is only one among several factors which do so. [pp. 145-148]

37. One of the most significant problems facing intervenor groups is that they rely, for the most part, on voluntary financial contributions. Difficulty in obtaining adequate funds to assume a full role throughout protracted licensing proceedings seriously limits the nature of citizen group input to the hearing process. It also creates an aura of uncertainty as to whether or not they will participate until a final decision is reached, a decision which may ultimately entail appeal to the courts. [pp. 194-195]

38. The cases we have examined lead to the conclusion that full-scale citizen interventions in opposition to construction or operation of a nuclear power plant cost in the neighborhood of $75,000-$100,000. The largest portion of the cost pays for lawyers fees and for the preparation of documentation. [p. 194]

39. Provision of adequate funds to citizen intervenors could, by removing the difficulties of logistics and the dependence on voluntary contributions of time and effort, eliminate the necessity for some considerable procedural delays. Such simple tasks as typing, duplication and distribution of briefs to all parties often strains the very limited financial and manpower resources of intervenor groups and is, in the course of an entire proceeding, expensive. [p. 195]

40. Citizen groups have often been accused of taking irresponsible stands, operating under the guise of the "public interest", without having to answer to any broader constituency. But the citizen groups with whom we have spoken claim only to represent their membership and many of the substantive positions they have taken in licensing and rule making hearings have been, even in the view of some applicant lawyers and many AEC officials, responsible. The concept of the "public interest" is an elusive one; in reality, any group represents only <u>its</u> interest and perhaps <u>its</u> view of the public interest. [pp. 202-204]

41. Intervenors are responsible to their own memberships in much the same way that utilities and manufacturers are responsible to their owners and stockholders. The representatives of intervenor groups taking part in hearings can no more afford to lose the confidence and support of their membership than the representatives of utilities and vendors can afford to lose the confidence and support of their management and stockholders. [p. 204]

42. Citizen groups (with the exception of certain individual members) make little effort to understand scientific and technological issues. They serve primarily as logistic supporters of the attorneys and participate little or not at all in developing the substance of the matters at issue. [pp. 187-204]

43. Since opposition interventions in individual nuclear power plant siting cases are essentially local, they are, in effect, one-time activities. In our experience there is no sophisticated or active informational network between or among various citizen group intervenors in different parts of the country. Thus, the experience of one group in mounting an intervention is likely to have little or no impact on other groups in other parts of the country. As suggested above, the only strong link among various citizen groups is the lawyer. [p. 23]

Science, Scientists, and Citizen Groups

44. Scientific advice to citizen groups is often obtained from graduate students or from individuals with discrete, often isolated, areas of expertise. The level of scientific and technical advice to intervenor groups is at best marginally adequate. [p. 205]

45. One result of the inadequacy of scientific advice to intervenors has been their reliance, particularly in individual plant hearings, on cross-examination to build their case. Attorneys and citizen groups would prefer to make a stronger presentation based on direct testimony and some even state that it is impossible to make a viable detailed scientific or technological case (one which will significantly affect the decision) by cross-examination alone. [p. 206]

46. Scientific help obtained by citizen groups is usually inadequate for several reasons: good scientific assistance is costly and virtually all citizen groups lack adequate resources to support a full scale scientific and technological analysis; it is difficult to find experts on nuclear power who do not have close ties with the nuclear industry or the AEC since government funding for nuclear power research comes, for the most part, from a single source; and, for the most part, witnesses and scientific advisors for intervenor groups act as individuals rather than as members of interdisciplinary or multidisciplinary teams. But consideration of matters involving the application of technology and an understanding of the effects of the use of technology are essentially interdisciplinary in nature,requiring a broad integrated team approach. Citizen groups frequently lack the necessary information and experience to recognize what kind of scientific help they need and where to locate it. [pp. 207-214]

47. The exclusivity of AEC sponsorship of reactor research and development is a basic and perhaps insurmountable obstacle to free and open debate within the scientific community and therefore, in the public arena, on matters involving safety and other aspects of nuclear technology. The AEC's role as the major and, in some particular areas, the exclusive funding source for research in the nuclear sciences has created a situation in which almost every nuclear scientist or engineer is under grant or contract to the AEC, or in the employ of the AEC, the nuclear industry, or a national laboratory. [pp. 210-211]

48. Scientific and technical details of nuclear power plants are not interesting or exciting to most people, with the possible exception of nuclear scientists and engineers. Thus, when the discussion turns to technical details, hearings lose any excitement which they might have for laymen. [pp. 212-213]

49. Problems involving the definition of the word "expert" have led to considerable procedural difficulties, particularly in the emergency core cooling system hearing. Concern about the qualifications of witnesses or technical interrogators has often taken precedence over the evaluation of the weight of their testimony or the value of their participation. [pp. 215-218]

50. "Adversary science", both as defined by many of its advocates and as practiced in AEC licensing proceedings, involves the use of that scientific and technical information which supports a particular case and the discarding or obfuscation of information which detracts from the strength of that case. [p. 222]

51. Adversary proceedings are not an efficient way to resolve issues involving scientific and technical details or to resolve scientific or technological issues. [pp. 220-222]

52. There is, at present, no mechanism for independent study of the scientific and technological issues involved in nuclear power cases. Such independent study could serve as an important input to a decision process which weighs the beneficial and harmful impacts of nuclear power plants. [pp. 222-223]

53. Determining the impact of a technology on human and societal values entails, as one important element, a definitive and accurate analysis of the impacts of the technology upon the physical and natural surroundings. But such a determination involves, as well, the consideration of more difficult, though no less important, factors less amenable to quantification such as quality of life, aesthetics, etc. [p. 218]

Other Parties to the Hearings

54. The strength of an intervenor's case often depends upon the adequacy and accuracy of the information provided by the applicant for a permit or license or, as in the case of the ECCS hearings, by the vendors. [p. 224]

55. Utilities and vendors have available to them the scientific and technological manpower and information resources necessary to make a comprehensive case and can utilize the services of large and sometimes talented groups of witnesses. [p. 229]

56. Vendors are generally reluctant to provide information to intervenors. Claims of proprietary rights frequently are asserted in order to justify the withholding of documents which is, in many cases, an additional factor leading to delay. [p. 225]

57. When utilities and vendors, with direct economic interests in the outcome of a proceeding, play the role of primary information source for intervenors, there is a potential for distortion and misrepresentation of information. [pp. 225-228]

58. The argument frequently made by AEC Staff, applicants and vendors, that complex technical matters cannot effectively or meaningfully be explained to laymen, is invalid; while the task if difficult, it must be recognized that the aim is not to make the public conversant about the details of nuclear power plant design. Rather, it is to explain the outstanding issues in a manner which makes it possible for the public, and other non-scientists, including governmental decisionmakers, to evaluate the importance of scientific and technical issues and to use the information as an input to the decisionmaking process. The public has a right to know, and the agency and applicants have a responsibility to inform them, particularly since the full costs of nuclear power development have been borne by the public. [pp. 228-229]

59. The straightforward and understandable presentation of those issues is a responsibility which all parties in the proceedings must share, a responsibility that, with very few exceptions, none have accepted. [p. 229]

Role of the AEC

60. The AEC is faced with the dual responsibility of promoting the development of nuclear power and of regulating that development to insure public health and safety. This has caused difficulties, both apparent and real, not only for the agency but for all of the parties involved in the proceedings. There are indications that a change in the structure of AEC and, in fact, of the nation's entire energy policy mechanism, is imminent. [pp. 230-235]

61. By the time the consideration of a permit or license application, or of a proposed rule, reaches the stage at which the public becomes involved, the AEC's Regulatory Staff has determined that the permit or license should be granted or that the proposed rule should be adopted. Thus, at the hearing,

the AEC Staff is, in effect, the ally of the applicant and, on virtually any question of import, the Staff's position is identical to the applicant's. The AEC is, then, an advocate in an adjudicatory proceeding in which it is also the judge. To the extent that the hearings remain an AEC instrument, they will be carried out in such a way as to result in the fulfillment of the Agency's promotional mission. [pp. 235-237]

62. The environmental impact statements filed by the AEC under the requirements of the National Environmental Policy Act provide a potentially valuable source of information for all parties involved in AEC proceedings. [p. 238]

63. Much of the information relied upon in the preparation of environmental impact statements is provided by applicants and reviewed by National Laboratories under contract to the Atomic Energy Commission. [p. 238]

64. Environmental Impact Statements have failed as information sources for citizens because of the bias inherent in their preparation, the overwhelming concern with procedural, as opposed to substantive characteristics of the statements, and the emphasis on the physical as opposed to the social and human environment. [pp. 238-242]

General

65. Much of the effort in AEC hearings is designed to create a record which is expected to serve different ends for each of the parties. The AEC seeks a record which will substantiate the granting of a license (or in rule making hearings, the formulation of criteria); the utilities hope the record will demonstrate the adequacy of their plans and support their application

(in rule making, vendors hope the record will show that criteria according to which they have designed their equipment and which they may already consider overly stringent should not be made more so); the AEC Staff wants the record to show that it has already, prior to the hearing, extracted sufficiently stringent guarantees for the proper construction and safe operation of the plant (in rule making hearings, the Staff hopes the record will substantiate its view of reasonable criteria; criteria not substantially different from those already existing); and intervenors hope the record will show the shortcomings of plans for a plant, the missing safeguards and potential dangers, the adverse impact on the physical and human environment, the cupidity of the utilities, evidence of procedural error upon which litigation can be brought, and a basis upon which information can be generated for public education (in rule making hearings the intervenor hopes the record will provide a basis for public education as well as for the most stringent criteria possible and provide a basis for subsequent court review).

66. Intervenors lack sufficient scientific and technological assistance, financial resources, data, analytical capabilities, and, perhaps most importantly, objective judges for their case. This puts them at a distinct disadvantage in carrying out their role in an intervention.

67. Despite lip-service paid to their importance in the process, intervernors are viewed by all other parties as interlopers who disrupt the orderly procedure of licensing plants.

68. Despite somewhat differing viewpoints, there is a common set of interests among the Atomic Safety and Licensing Boards, the AEC Staff, utilities, vendors (and, at times, primary economic beneficiaries) which effectively makes them allies in licensing proceedings.

69. The sheer volume of documents and materials and their organization in the AEC public document rooms tends to obfuscate essential matters in contention.

70. The decision of ASLB's to recommend the licensing of plants seems to be only incidentally, if at all, related to the resolution of issues raised by intervenors during hearings.

71. Scientific and technological assessments must be viewed as an input to the decisionmaking process rather than as the decision itself.

RECOMMENDATIONS

The recommendations included in this section of the report were developed as a result of our detailed study of the participation of citizen groups in the licensing and rule making proceedings of the Atomic Energy Commission. Some relate directly to these proceedings; others are more general and can be applied to the participation of citizen groups which direct their efforts to other environmental issues in other forums. The applicability of some are distinctly dependent on the adoption of others and, therefore, must be considered in context.

We have found that the AEC proceedings fail to provide a meaningful forum for public participation. We have also found that citizen group use of scientific and technological resources is seriously deficient. The recommendations which follow represent an attempt to devise mechanisms which would provide a means of achieving a greater public understanding of scientific and technological issues, which would ensure the fair and judicious consideration of matters of concern to the public and which would allow for more adequate public participation in the decisionmaking process.

Some of the recommendations, if implemented, would involve major and significant institutional change but it is our view that no amount of procedural adjustment or minor changes in the process will result in full, fair, independent, and judicious consideration of the issues placed in contention by citizen group intervenors. Small changes can only result in affording minor conveniences but will not significantly alter a process in need of serious revision and redirection. In our view, at least, no system will work without a major attitudinal change on the part of legislators and officials of the Executive branch of the Federal government. Clearly, a significant part of the Congress has become concerned about the effects of technology and the need for greater citizen participation. Ample evidence to demonstrate the concern can be found in legislation enacted over the last several years. The extent to which the Congress or the Executive branch will actually implement recommendations for significant changes in the process of citizen participation, of course, remains to be seen.

We feel that reemphasizing more active citizen involvement in governmental decisionmaking is essential, particularly in areas where such decisions entail the division of resources which can be expected to have a significant impact on the community. Citizen dissatisfaction with the ways in which matters involving the environment, both physical and human, have been resolved in the past has become evident in recent years. Antipathy to elitist governmental and private sector decisionmaking processes is more obvious and has manifested itself in the intensity and frequency of citizen protest activities.

-- Since one of the critical problems in citizen group use of scientific and technological resources is the lack of independent and impartial evaluation of the technical issues involved in nuclear power cases (as well as in many other matters involving the application of technology) [Findings 4, 15, 44, 46, 47, 51, 52],

> 1. We recommend the formation of independent assessment centers to perform and arrange for analyses of the impact of technology on the physical, social and human environment.

Such centers would operate independent of any mission-oriented agency and would form a nucleus for carrying out <u>ad hoc</u> interdisciplinary studies of the consequences of actions involving the use of technology. Supported by Congressional appropriations and by State contributions, the centers would thus provide an input to the decisionmaking process which is presently non-existent. Rather than being structured along traditional geographical lines (by state or region), the centers could be organized about areas of similar environmental characteristics or problems. Thus, areas of the country with similar problems (water resources, air quality, land use, urbanization, etc., for example) could be joined in a single center independent of their geographic location. The staffs of the centers would assemble assessment groups to perform analyses of particular environmental problems. Groups performing assessments would be comprised of experts from a broad variety of disciplines relevant to the problem being assessed and would be on reimbursable loan from the teaching and research faculties of universities and from research institutions for the duration of a particular study. In order to avoid the development of entrenched and bureaucratic staffs with prevailing attitudes or positions within the centers, the staffs of the centers should serve with limited tenure for a specific term, perhaps three years, and should then be assisted by the centers in locating new positions. It is probable that the value of experience gained in such a position would be of considerable import to government agencies, universities and private industry. The studies carried out by the *ad hoc* interdisciplinary teams, in addition to providing necessary information to decisionmakers and the public, would, if the centers and their teams were maintained scrupulously independent of any vested interests, avoid the now prevalent problem of continuous and acrimonious conflict over the motives of scientists and engineers providing information to policymakers (for example, about the effects of nuclear power plants or other technologies). The centers could each interact closely with each other and with the Congressional

Office of Technology Assessment, when that Office begins operating. Interdisciplinary analyses performed by the centers would integrate the best available expertise and knowledge about the human and physical environment and other effects of a technology, and, perhaps, initiate new investigations in areas where definitive information is lacking. The analyses would clarify areas of agreement among experts; they would, as well, spell out areas of disagreement or uncertainty so that decisionmakers, in weighing the ramifications of various alternative actions, could base their decisions upon an understanding of the state of knowledge concerning the effects of those actions. It must be emphasized that the assessment centers would make no decisions but would provide information to the decisionmakers about the known and potential impacts of technology. Analyses would be made available to the public and would thus provide a definitive basis for citizen group involvement to supplement what is, at present, a largely visceral basis for such involvement.

The need for environmental assessment centers has been recognized in a bill which would amend the National Environmental Policy Act to encourage the establishment of state and regional environmental centers which is presently pending before the Senate Committee on Interior and Insular Affairs (S.1865). The purpose of the legislation is to provide for and supplement existing programs for the conduct of basic and applied research, investigations and experiments relating to the environment. The Administrator of the EPA would be authorized to distribute $10 million annually through FY 1976 on the basis of population, total land area, the nature and severity of the environmental problems in the areas served by the centers, and the ability and willingness of each center to address these problems. Some of this money would be only available to those centers for which the concerned States themselves provide the requisite matching funds.

The legislation holds that the centers would have "a nucleus of administrative, professional, scientific, technical, and other personnel capable of planning, coordinating and directing interdisciplinary programs. . . ."[1] The proposed legislation is similar to what has been proposed here although it places greater emphasis on traditional state and regional boundaries than we feel is necessary or advisable in the examination of environmental matters.

-- We state elsewhere in this report that the current hearing process is marked by a lack of a fair and unbiased adjudicatory body [Findings 26, 27, 29, 60, 68]. Accordingly,

> 2. We recommend the formation of governmental bodies of adjudicators independent of any mission-oriented agency to hear and rule on the arguments of competing interests in nuclear power licensing cases.

Bodies of this sort, independent of any mission-oriented agency, would serve as a mechanism for the consideration of licensing of nuclear power plants, though there is clearly implication for the adjudication of other types of environmental or technology based disputes. The adjudicators would make recommendations to the licensing authority; the licensing authority would not be legally bound to follow the recommendations of the adjudicators, although, if it ignored the recommendations, it would be required to explain, in detail, the reasons for its action and to demonstrate that it was not acting in an arbitrary or capricious manner. The body of adjudicators could be empowered by legislation to serve more than one Executive branch agency. As with the assessment

----

[1] S.1865, 93rd Congress, 1st Session, p. 7, lines 7-10.

centers, except for a skeletal staff of managers, there would be no long term appointments to this body of adjudicators, in order to avoid the formation of any set and definitive opinions about the issues to be considered. In any given case, three judges, including two with relevant scientific or technical competence and as is presently the case with ASLB's, one with legal competence, would be appointed to preside over a proceeding designed to fairly and judiciously consider issues raised by all of the parties involved and to raise any questions or issues which it felt were not adequately posed by the parties. Adjudicators would be chosen from a panel in much the same way that ASLB members are chosen; the significant difference from current practice would be that selection would be within a mechanism with no relationship to the AEC or any of the parties involved in the case.

There are, of course, precedents for such independent adjudicative bodies within the Federal structure. Both the Federal Mediation and Conciliation Service and the National Mediation Board, which mediate labor disputes, contain organizational elements applicable to this concept. Decisions of the adjudicators, of course, would be subject to court review.

-- The impact of the dual role of the Atomic Energy Commission on the fairness of the hearing process has been treated in this report [Finding 60] and elsewhere;

> 3. We recommend that the developmental and regulatory roles of the Atomic Energy Commission be separated by the formation of two independent agencies, one concerned with the development of nuclear energy (and perhaps with other sources of energy) and the other concerned solely with the regulation of nuclear energy (and perhaps other sources of energy) for the purpose of protecting the public health and safety.

Several current proposals have the effect of splitting the Atomic Energy Commission's functions although at the time of this writing, none have as yet been acted upon. Prominent among these suggestions is the President's proposal to create an Energy Research and Development Administration (ERDA) as part of a reorganization of Federal agencies which share the responsibility for energy policy. According to the pending legislation now before the Senate and House Committees on Government Operations, it is in the public interest for the licensing and related regulatory functions of the AEC to be performed separately from the other functions of the Commission.[2] The legislation would transfer the promotional programs of the AEC to the newly created ERDA, while the licensing and related regulatory functions of the Commission would be carried out by the renamed Nuclear Energy Commission. We believe that this proposal would be strengthened were it amended to include independent adjudicatory bodies as recommended above (Recommendation 2). The existence of a separate regulatory agency would permit the participation of a Regulatory Staff without the appearance, or reality, of bias, in the process of evaluating applications for permits and licenses.

-- The preceding three recommendations involve major changes in institutional structure; only the third has been publicly discussed seriously enough to realistically anticipate action on it in the near future. The other proposals would require considerable discussion and evaluation within the Congress and the Executive branch, and by the public. Major institutional changes are

----

[2] Hearings on S.2135 before the Senate Committee on Government Operations, Subcommittee on Reorganization, Research and International Organizations, August 1, 1973.

not easily implemented; this causes us some concern as we are aware that the viability and usefulness of our recommendations depends upon their chances for eventual adoption. Yet the problem of citizen involvement in decisions about the use of nuclear power and technology in general seems to require at least these major changes. We feel that these recommendations are realistic, although clearly difficult to implement.

The remainder of our recommendations require less far-reaching changes in the existing system, but, we believe, they are critically important for the realization of a meaningful process of citizen participation in these matters.

-- We have pointed out elsewhere in this report that citizen group frustration and impotence at AEC hearings arises in part from their late entry into the consideration of issues [Findings 15, 61]. Accordingly,

> 4. We recommend that the public be informed of any plans for the construction of a nuclear power plant at the earliest time possible and no fewer than five years in advance of the planned beginning of construction. Government consideration of an application for construction of a nuclear power plant should provide for citizen group opponents to become involved at the time that the preliminary application for a permit (or license) is first submitted.

Involvement at the earliest possible stage, such as when a utility begins to consider the need for a new generating facility, would be helpful since it would allow citizens to have a voice from the very start in planning future development. If citizens and citizen groups played a role in early planning and consideration of nuclear power plants, it is likely that they would participate in a less negative manner than they do now. It must be emphasized that the opportunity for citizen participation must be real and not token; citizen concerns must be faced squarely, honestly and judiciously and citizen opinions taken seriously. All parties would benefit; utilities would no longer face long, contentious, and acrimonious proceedings or uncertainty about investments, and citizens would be able to make their voices heard.

A number of bills currently under consideration by the Congress would have the effect of implementing all or part of this recommendation. For example, the proposed "Power Plant Siting Act of 1973" (H.R. 3591) requires utilities to prepare long range plans for bulk power supply facilities which include, among other requirements, the identification of "tentative sites for the construction of future powerplants . . . including an inventory of sites for all plants on which construction is to be commenced in the succeeding five years. . . ."[3] The bill also provides for the establishment of State or regional certifying bodies which, in their operation, "may also provide for participation by members of the public."[4] A bill introduced by Congressman Dingell (H.R. 180) contains similar provisions with an alteration which requires that State or regional certifying bodies "_shall_ also provide for participation by members of the public."[5]

-- Holding two sets of hearings on the same set of issues has distressed all parties involved in the nuclear power licensing process [Findings 8, 9, 11]. Construction permit and operating licensing hearings seem essentially duplicative; however,

----

[3] H.R. 359, 93rd Congress, 1st Session, Sec. 4.(a)(2), p. 6, lines 25 and p. 7, lines 1-4.

[4] _Ibid_., Sec. 5(a)(1), p. 9, lines 1-2.

[5] H.R. 180, 93rd Congress, 1st Session, Sec. 5(a), p. 10, lines 5-6.

> 5. We recommend required public hearings at both the construction permit and operating license stages. The construction permit hearing should consider all issues relating to radiological safety and environmental impact. The operating license hearing should determine only whether the requirements of the construction permit have been adhered to (and should consider new information about safety, environmental impact, quality control, or deviations from approved plans which has become available only after the construction permit hearing).

The construction permit hearing would, in most cases, be the more significant of the two hearings since it would be at that stage that major safety and environmental problems would be considered and resolved. But the purpose of dual hearings would be to provide a mechanism for the organized examination of the extent to which requirements and commitments, incurred during construction permit proceedings, are fulfilled. Currently, the follow-up is performed out of view of the public by the Regulatory Staff. With follow-up as the primary concern of a set of hearings, public participation in the determination of the extent of utility adherence with construction permit requirements would be guaranteed. Agency quality control inspectors, should, as well, be called upon to testify at the second set of hearings in detail, be required to submit substantiating evidence, be subject to cross-examination and to certify their findings of compliance or non-compliance. Of course, for such a procedure to be effective, all requirements and commitments would have to be carefully and explicitly spelled out in the construction permit; the operating license hearings could also consider new evidence. But since it is clearly both expeditious and advantageous to all parties concerned to have as many issues as possible raised at the earlier construction permit phase, adjudicators would have to be strict in their interpretation of "new" evidence to avoid a *de novo* review of all safety and environmental issues considered earlier.

-- As utility companies continue to propose more nuclear power plants, it becomes clear that certain issues common to all or most plants are being raised in the consideration of applications for individual construction permits and operating licenses. Clearly such a situation is not administratively expeditious and leads to the possibility of conflicting decisions on identical matters [Finding 10]. To avoid unnecessary duplication and to improve the chances of judicious, and yet consistent, consideration of generic issues,

> 6. We recommend increased use of generic hearings to consider issues common to all or most nuclear power plants. At the same time, steps must be taken to encourage public participation in these generic hearings.

A caveat in this recommendation is necessary because, as we have pointed out elsewhere in our report, citizen involvement is, for the most part, local in nature. Generic hearings are by necessity not local, but are rather national in scope, so that a particular effort must be made to explain publicly and relate generic issues to individual plants so that local public interest can be maintained and concerns allayed. Generic hearings should possess all of the characteristics of adjudicatory hearings: discovery, direct testimony, cross-examination, etc. Arguments that adjudicatory hearings on generic issues encourage delay ignore the fact that even the overlong ECCS proceeding is administratively more expeditious than would have been the consideration of emergency core cooling system criteria at every individual plant licensing hearing held throughout the country.

It should also be noted that a system of generic hearings would be more effective if greater standardization of nuclear power plants were the rule. As long as each nuclear power plant is, to all intents and purposes, a unique entity, many issues will remain outside the scope of any generic proceeding.

-- The local nature of citizen group interventions in nuclear power cases has led to costly and inefficient duplication of effort by citizen groups. No regular effective mechanism presently exists for the exchange of information among various groups, with the exception of the input made by intervenor lawyers serving more than one citizen intervenor group [Findings 43, 46]. Because of the limited funds, it would be helpful if new groups could learn from the mistakes and successes experienced by existing organizations. Therefore,

> 7. We recommend that a mechanism be established (perhaps with private foundation funding) to permit the exchange of information among citizen groups about substantive issues and about the process, techniques, tactics, and strategies of public involvement.

Such a mechanism could provide information, for example, about scientists, engineers, and lawyers who are willing and able to provide assistance to citizen groups; about documents, articles, or books and other information sources which have proven useful in the past; about potential sources of funds; about methods and techniques of organization and public education; etc. There is no need for citizens to begin their involvement in these matters as though they are the first to attempt to take part in a governmental decisionmaking process. They should be able to draw on the experience of others so that they can make the best possible use of their severely limited funds and maximize the effectiveness of their participation.

-- Citizen groups find themselves at a great disadvantage in licensing hearings because of the difficulty they have in raising funds and because of the disparity in financial and manpower resources between these groups and the proponents of nuclear power plants. The paucity of funds manifests itself in the weakness of intervenors' case and often in the procedural delay of hearings for reasons of simple logistics. The level of debate would clearly be raised, and the grounds for delay and procedural haggling reduced, if citizen groups were better funded [Findings 7, 37, 38, 39, 46, 66]. We have stated elsewhere that in a democracy there is an obligation to maintain decisionmaking processes open to citizen involvement and input. Considering and granting construction permits or operating licenses for nuclear power plants represents an area where at least some citizens have demonstrated a willingness and desire to make the necessary sacrifices to be involved. That involvement, we believe, should be maximized and made meaningful. Therefore,

> 8. We recommend that legitimate citizen group interventions in opposition to the granting of a permit or license or in the consideration of a proposal for a rule change be funded through the assessment of an application fee paid by the applicant for a permit or license.

Such a measure would assure citizen involvement in the licensing process and indeed enhance it. To the advantage of all parties it would, moreover, eliminate the excuses of poverty which presently justify delay and deviations from rules or other requirements. We are, of course, aware that the adoption of this recommendation would raise certain difficult issues. For example, what constitutes a "legitimate citizen group intervention"? A clear definition would have to be formulated and interpreted by the adjudicatory body hearing the case. Such a definition, as we have pointed out in our report already exists and such determinations are now made using the standard by which parties are now granted leave to intervene before the ASLB.[6] Making such a determination would be less open to criticism if Recommendation 2 (the formation of bodies of independent adjudicators) were implemented; if the AEC were to maintain control over the hearing process, determinations of standing for the purpose of

----

[6] 10 CFR §2.714.

receiving funding would be less satisfactory and would likely lead to interminable wrangling among competing parties. In our view, some independent body would have to be established to make determinations concerning "legitimate interventions." The independent adjudicators (or another independent body) would also make decisions about the advisability of consolidating interventions by separate citizen groups if it were determined that the goals and contentions of the groups were similar enough. (An alternative would be to permit only a single intervention by citizen groups opposed to the granting of a permit or license and to require any citizen groups desiring to take part in opposition interventions to make agreements among themselves about the use of funds which would be provided if this recommendation were adopted.) Clearly, considerable discussion among the various parties to AEC proceedings would be useful in arriving at a clear definition of the phrase "legitimate intervention" and an agreement about the conditions under which separate interventions might be consolidated. Our recommendation would, we realize, raise additional problems, but we feel that these are not insurmountable. Interventions, although expensive (at present, $75,000-$100,000 for a full-scale intervention)[7] still constitute an extremely small fraction of the total costs of constructing a nuclear power plant, so that the adoption of this recommendation would not result in significant financial hardship for the applicant. This recommendation recognizes the importance to all parties to the proceedings of

---

[7] The cost of a full-scale intervention under our proposal might be higher or lower, depending on whether or not our recommendation #1 were adopted. If it were, the cost to the intervenors might be lower as at least part of the amount they now spend for scientific and technical help would be reduced (to them) but would be borne by the assessment center; if it were not adopted, considerably more would need to be expended, in our view, on adequate scientific and technical assistance for intervenor groups.

full and meaningful citizen participation and places the financial burden for this participation upon the party most able to meet the cost, the party proposing a change in the status quo, and, incidentally, the party likely to profit economically from the proposed action. (It might be suggested that, in the course of a complete licensing proceeding, the adoption of this recommendation might actually reduce the cost to the applicant by eliminating unnecessary delay in the total process.)

-- But a broader public than merely the citizen groups and their members should have access to the licensing process. At the very least, the public-at-large should be made cognizant of the issues under contention and be encouraged to attend hearings. Presently, hearings are held during the day on working days; thus, attendance is usually limited to parties with a direct interest (primarily the applicant utility's employees, in most cases) and to those few members of the public able to afford the time [Findings 1, 30, 34]. To remedy this,

> 9. We recommend that public hearings be made accessible to a broader public by requiring evening or weekend sessions to the extent possible and by encouraging the use of community cable and public television or radio for local broadcasting of hearings.

Further, since the subject matter of licensing hearings is often complex and detailed and, as currently presented, beyond the understanding of lawyers, Board chairmen and the public [Findings 18, 19, 48, 58],

> 10. We recommend that all participants in public hearings be required by the presiding officer to present summaries of all formal statements and written comments (no matter how complex) in language understandable to educated laymen (and in particular to non-scientists, non-engineers, and non-lawyers).

It is true that the argument is often made that matters of technology are so complicated that it would be futile to attempt to make them understandable to the lay public. But it is our experience that scientists and engineers with a true understanding of their own endeavors can, and have been able to, explain their work to laymen. This is not to say that the scientists or engineers can turn the public into instant experts, but they can, and we argue should, apprise the public of the nature and essence of their testimony in a clear and understandable manner. As we view the implementation of this recommendation, the adjudicatory Board would require the preparation by technical witnesses of abstracts of their testimony which are translated into clear and concise English prose.

The Board could, at its own discretion, remand any such presentation to the witness as not sufficiently clear and understandable, and require further clarification to its satisfaction. The object, of course, would be the explication of testimony, not necessarily to its finest detail but most certainly adequately enough to explain the testimony and the basis for it to an interested, if not technically trained public. Copies of such abstracts should be made available to the public at the hearings.

During cross-examination of witnesses the Board could similarly require explanation or clarification of testimony for the purpose of public understanding.

It is not an easy task, but it is by no means impossible. Each year, for instance, Congress considers our huge national scientific and technological enterprise and appropriates the funds necessary to support it. In our experience, no agency of government testifies before Congressional committees that the issues are too complicated for committee members, as laymen, to understand. Even complex matters relating to research and development programs and expenditures are probed in detail by the Congress, and agency officials take great pains to make their testimony clearly understood. No less can or should be done for the public.

We have noted throughout our study the repeated contention of citizen group attorneys that the unavailability of certain documents presents difficulties in the structuring of a meaningful position. Frequently, as we have mentioned, utilities and vendors withhold some documents under the claim of proprietary rights. At times, such claims are, undoubtedly, legitimate; at other times, they are apparently specious. The necessity for private companies to maintain exclusive rights over equipment design, construction techniques, or other trade practices clearly conflicts with the necessity that the public be aware of the impact of the technology upon its health, welfare and safety [Findings 13, 15, 54, 56, 59]. This conflict, while present in the development of other technologies as well, is particularly important in matters of nuclear power, a technology developed at public expense and with such great implications for health and safety. In an attempt to resolve the conflict,

> 11. We suggest that legislation be designed and considered by the Congress to institute a process of licensing of trade practices, equipment design, and construction techniques for the exclusive utilization of the company responsible for their development. All relevant documents, then, would be made a part of the public record.

The documents would be available for the use of citizen groups and any other interested parties. Documents relating to developments which were publicly funded would be automatically available. The licensing of other techniques or practices would require full disclosure, but would protect the interests of individual companies. Needless to say, this suggestion has broad and serious implications for patents, copyrights, international agreements, etc. We would, of necessity, then pose it in such a manner so as to encourage its consideration and hopefully, a full exploration of the viability of this concept by the Congress. Admittedly we have not gone through that process and recognize that it must and should be dealt with more thoroughly than we are able to do at this time. Rather than a recommendation, we pose it simply as an idea seemingly worthy of further elaboration, exploration and consideration.

APPENDIX A: LIST OF INTERVIEWS

During the course of this study, we interviewed over 100 persons whose names appear below. In those cases in which no date is given, several interviews or discussions took place over an extended period of time.

Midland

Ned S. Arbury	(Midland Nuclear Power Committee)	May 19, 1972
Robert Asperger	(Saginaw Intervenors)	May 19, 1972
Myron Cherry	(Attorney, Saginaw Intervenors)	
Steve Gadler	(Advisor, Mapleton Intervenor)	May 17, 1972
Clark Goodman	(ASLB)	May 18, 1972
David Hall	(ASLB)	May 18, 1972
Louis Hamilton	(University of Michigan)	June 1, 1972
Larry Holcomb	(Creighton University)	May 17, 1972
David Kartalia	(AEC)	May 18, 1972
Irving Like	(Reilly, Like, Schneider; Attorney, Mapleton Intervenors)	May 17, 1972
Wendell Marshall	(Mapleton Intervenors)	May 18, 1972
Ken McCue	(Purdue University)	November 1, 1972
Arthur Murphy	(ASLB)	May 18, 1972
Wayne North	(First United Methodist Church of Midland)	June 7, 1972
Claudia Protz	(Consolidated National Intervenors)	
Mary Sinclair	(Saginaw Intervenors)	May 17, 18, 1972
Thomas Slykhouse	(Saginaw Intervenors)	May 18, 19, 1972
Phyllis Slykhouse	(Saginaw Intervenors)	May 18, 19, 1972
Stan Wiggins	(University of Michigan)	May 17, 1972

Vermont Yankee

Beatrice J. Brown	(Brattleboro resident)	August 15, 1972
J. Wilcox Brown	(NECNP)	September 30, 1972
Jane Brower	(Mt. Holyoke College)	September 30, 1972
John Calhoun	(State of Vermont)	August 15, 1972
Harvey Carter	(Williams, Witten, Carter, and Dollard)	August 16, 1972
Albert Cree	(Vermont Yankee Nuclear Power Corporation)	May 31, 1972
Quint Dawson	(NECNP)	September 30, 1972
Thomas Dignan	(Ropes and Gray; Attorney, VYNPC)	
Marc Goldsmith	(MIT)	June 20, 1972
Mark Haflich	(AEC)	June 1, 1972
Everett Hafner	(Hampshire College)	June 21, 1972
Allan Hoffman	(University of Massachusetts)	June 20, 1972
Brendan Hutchinson	(Cornell University)	August 16, 1972
David Inglis	(University of Massachusetts)	June 20, 1972
Gerald James	(Conservation Society of Southern Vermont)	May 31, 1972

Vermont Yankee (cont'd.)

Lawrence Keyes	(Vermont Yankee Nuclear Power Corporation)	June 1, 1972
Richard Kofler	(University of Massachusetts)	June 2, 1972
Harriet Lyon	(NECNP)	June 1, 1972
Irving Lyon	(NECNP)	June 1, 1972
William Massar	(AEC)	
Esther Poneck	(NECNP)	
John Ritsher	(Ropes and Gray; Attorney, VYNPC)	June 2, 1972
Anthony Roisman	(Berlin, Roisman and Kessler; Attorney, NECNP)	
Norman Runnion	(<u>Brattleboro Daily Reformer</u>)	June 1, 1972
Karin P. Sheldon	(Berlin, Roisman and Kessler; Attorney, NECNP)	
Diana Sidebotham	(NECNP)	
Nathaniel Smith	(NECNP)	May 31, 1972
Polly (Plimpton) Smith	(NECNP)	May 31, 1972
Donald Stever	(State of New Hampshire)	June 1, August 15, 1972

ECCS

Thomas Arnold	(Epstein, Salloway and Kaplan; Attorney, CNI)	September 21, 1972
Bart Cowan	(Eckert, Seamans, Cherin and Mellott; Attorney, Westinghouse)	
George Edgar	(Morgan, Lewis and Bockins; Attorney, Babcock and Wilcox)	
Daniel Ford	(Union of Concerned Scientists)	
George Freeman	(Hunton, Williams, Gray and Gibson; Attorney, ECCS Utility Group)	
Nathaniel Goodrich	(ASLB)	September 22, 1972
Henry Kendall	(MIT; Union of Concerned Scientists)	
James McKenzie	(National Advisory Society; Union of Concerned Scientists)	June 20, 1972

Other

Richard Balzhiser	(OST)	October 11, 1972
Philip Bereano	(Cornell University)	
Richard Berg	(Administrative Conference of the United States)	September 20, 1972
Graham Chadd	(AAAS)	April 11, 1973
David Comey	(Businessmen for the Public Interest)	September 7, 1972
Robert Crawford	(HEW)	June 23, 1972
Paul Cromwell	(HEW)	June 23, 1972
Edward David	(President's Science Advisor)	October 11, 1972
Warren Donnelly	(Congressional Research Service)	
William Doub	(AEC)	November 27, 1972
Leonard Dworsky	(Cornell University)	

Other (cont'd.)

Kay Eckert	(AAAS)	April 11, 1973
Charles Ehren	(University of Denver)	September 23, 1972
Thomas Engelhardt	(AEC)	May 2, 1972
James Fay	(MIT)	June 19, 1972
Muriel Ferris	(Administrative Assistant, Senator Hart)	May 8, 1972
Jerome Garfinkel	(ASLB)	
John Gofman	(University of California, Berkeley)	May 16, 1973
Harold Green	(George Washington University)	
Rainer Gross	(Federal Ministry for Education and Science, Germany)	September 23, 1972
Bernard Hartung	(Senate Subcommittee on National Parks)	April 3, 1973
F. Hetman	(OECD)	May 30, 1973
Martin Hoffman	(AEC)	October 3, 1972
Michael Joy	(Administrative Assistant, Senator Hollings)	April 3, 1973
Gar Kaganovich	(Executive Assistant, Senator Case)	
Peggy Kennes	(Kettering Foundation)	April 3, 1973
Alan Kulakow	(EPA)	
Ralph Lapp	(Quadri Science, Inc.)	
Angus MacBeth	(Natural Resources Defense Council)	July 17, 1972
James Mahoney	(President's Council on Management Improvement)	
John McCarthy	(Department of Justice)	
John McConnell	(N.Y. Stock Exchange)	
Mace Miyasaka	(Johns Hopkins University Applied Physics Laboratory - Maryland State Study on Power Plant Siting)	September 27, 1972
George Murphy	(Joint Committee on Atomic Energy)	September 7, 1972
Dorothy Nelkin	(Cornell University)	
K. Oshima	(University of Tokyo)	May 30, 1973
William Painter	(Washington Ecology Center)	August 24, 1972
Skip Price	(EPA)	September 7, 1972
Stanley Prussin	(University of California, Berkeley)	May 16, 1972
James Ramey	(AEC)	November 6, 1972
Charles Reed	(National Academy of Sciences, Committee on Nuclear Physics)	
Robert Ritzmann	(AEC)	
Ronald Schiffman	(Pratt Institute)	
Miller Spangler	(AEC)	May 30, 1973
Carlos Stern	(EPA)	May 4, 1972
E. Winslow Turner	(Senate D.C. Committee)	
Paul Turner	(Atomic Industrial Forum)	July 7, 1972

APPENDIX B: THE CONSOLIDATED NATIONAL INTERVENORS

The Consolidated National Intervenors (CNI) at the ECCS hearings comprise the following groups and individuals:

Union of Concerned Scientists, Boston, Massachusetts

Sierra Club

United Auto Workers of America

Saginaw Valley Nuclear Study Group

Citizens Committee for the Environmental Protection of Michigan

Trout Unlimited

West Michigan Environmental Action Council, Inc.

University of Michigan Environmental Law Society

Businessmen for the Public Interest

Protect our Wisconsin Environmental Resources

Environmental Action, Inc., Hilton Head Island, So. Carolina

Survivors, Inc., Haledon, New Jersey

Citizens Committee for Radiation Control, New York, N.Y.

Horace E. Bailey

William F. Faust

Nelson Aeschliman

John L. Wentworth

Hilda M. Wentworth

John Liddell

Loretta Liddell

Wendell H. Marshall

Lloyd Harbor Study Group, Inc.

Federation of Homemakers, Arlington, Virginia

Wisconsin's Environmental Decade, Inc.

Community Action to Reverse Pollution, Gary, Indiana

Joint Schools Committee for Academic Excellence Now, Inc.
New York, New York

Conservation Society for Southern Vermont

Vermont Natural Resources Council

Lake Champlain Committee

Citizens League for Education About Nuclear-Energy, Inc.
New Rochelle, New York

Rockland Ecological Coalition, Inc., Spring Valley, N.Y.

Izaak Walton League of America, Chesterton, Indiana

Scientists and Engineers for Social and Political Action, NYC

Protect Your Environment, Albany, New York

4-H Earthkeepers, Croton-on-Hudson, New York

Carolina Environmental Study Group, Inc., Charlotte, N.C.

Oregon Environmental Council, Portland, Oregon

Eugene Future Power Committee, Inc., Eugene, Oregon

Citizens' Rights Committee, Hicksville, N.Y.

Independent Phi Beta Kappa Environmental Study Group, NYC

Pompeston Environmental Committee, Cinnaminson, N.J.

Rockland County Conservation Association, Inc., Spring Valley, N.Y.

Ecology Action of Oswego, Oswego, New York

Environment: Pittsburgh, Pittsburgh, Pa.

Society for Social Responsibility in Science, Pottstown, Pa.

Cortlandt Conservation Association, Inc., Croton-on-Hudson, N.Y.

Environmental Coalition on Nuclear Power, Norristown, Pa.

Delaware Valley Committee for Protection of the Environment,
Moorestown, N.J.

Citizens Committee for Protection of the Environment, Ossining, N.Y.

Anti-Pollution League, Allentown, N.J.

New England Coalition on Nuclear Pollution, Inc., Brattleboro, Vermont

APPENDIX C: SOME ARTICLES AND BOOKS ABOUT THE NUCLEAR POWER CONTROVERSY

Articles:

1. Abelson, P.H. "Nuclear Power; Rosy Optimism and Harsh Realities". Science, Volume 161 (July 12, 1968), 113.

2. "Another SST? Questioning the Development and Proliferation of the So-called Liquid Metal-Cooled Fast Breeder Reactor". Environment, Volume 13 (July 1971), 18-19.

3. "A-Plants: Wealth and Worry". New York Times, September 20, 1969, p. 30, Column 3.

4. "Atomic Accidents". Forbes, Volume 102 (November 15, 1968), 58.

5. "AEC: Commitments to Come". Science News, Volume 94 (September 28, 1968), 310.

6. "AEC Goes Public on Nuclear Safety". Business Week, (September 20, 1969), 52.

7. "AEC Moves Toward Sale of Gas Plants to Industry". Technology Week, Volume 20 (January 9, 1967), 26.

8. "AEC to Nuclear Industry: We're No Longer Your Protector". Science News, Volume 100 (October 30, 1971), 290.

9. Armagnac, A.P. "World's Biggest Atom Power Plant; Brown's Ferry Nuclear Plant". Popular Science, Volume 195 (September 1969), 94-7.

10. "Atom Power Plant in Hot Water; Florida Power and Light's Biscayne Bay". Business Week, (June 29, 1968), 69-70.

11. "Atoms Downtown; Questioning a Nuclear Plant in New York City". Time, Volume 82 (July 26, 1963), 42.

12. "Atomic Dangers Overrated". Science News Letter, Volume 86 (September 5, 1964), 149.

13. "Atomic Jitters; Proposed New Station at Corral Canyon, Los Angeles". Newsweek, Volume 62 (November 25, 1963), 73.

14. "Atomic Power in Industry: A Slump Now, But....". U.S. News and World Report, Volume 67 (July 7, 1969), 54-5.

15. "Atomic Power Plants for Your Town? Nuclear Electricity is Practical But There Are Problems". Changing Times, Volume 21 (April 1967), 39-41.

16. Bazell, R. J. "Reaction on Morningside Heights: Denial of Reactor Permit". Science, Volume 174 (November 5, 1971), 575.

17. Behrens, C. "Permit Process a Bottleneck, AEC Licensing of Reactors for Industry". Science News, Volume 92 (December 16, 1967), 594.

18. Betts, C.A. "Power to Burn; Peach Bottom, Pennsylvania". Science News, Volume 89 (June 4, 1966), 444-5.

19. Boffey, P.M. "Radioactive Pollution: Minnesota Finds AEC Standards Too Lax". Science, Volume 163 (March 7, 1969), 1043-4.

20. "Boil up for Superheaters; GE-AEG Joint Research Plan". Business Week, (May 23, 1964), 134.

21. "Bottleneck for the Nuclear Genie". Business Week, (December 5, 1970), 50.

22. Boyle, R.H. "Nukes Are in Hot Water; Thermal Pollution Caused by Hot Water From Nuclear Plants". Sports Illustrated, Volume 30 (January 20, 1969), 24-8.

23. "Build the First One in Washington". Christian Century, Volume 80 (December 4, 1963), 1491.

24. Burden, W.D. and Foster, H. "Those Atomic Power Plants". Excerpts from Letters, Living Wilderness, Volume 33 (Spring 1969), 28-30.

25. Cairns, J. "We're in Hot Water". Scientist and Citizen, Volume 10 (October 1968), 187-95.

26. "Cayuga's Waters; Thermal Pollution by a Proposed Power Plant". Nation, Volume 207 (December 30, 1968), 709.

27. Clapp, W.J. "Electric Power: New Uses, New Sources, More Demand". Interview, U.S. News and World Report, Volume 62 (March 27, 1967), 74-8.

28. Cloud, W. "Is Atomic Industry Risking Your Life?". Popular Science, Volume 186 (June 1965).

29. "Coal Power Vs. Nuclear". Scientific American, Volume 214 (February 1966), 50.

30. Commoner, B. "Myth of Omnipotence; the Hidden Costs of Nuclear Power". Environment, Volume 11 (March 1969), 8-13.

31. "Con Ed Loses Its Nuclear Punch". Business Week, (July 4, 1970), 19.

32. "Cooling It In Minnesota". Report of the Minnesota Committee for Environmental Information on the Benefits and Risks of Nuclear Reactors, Environment, Volume 11 (May 1969), 21-5.

33. "Cooling Threat to Nuclear Power". Business Week, (June 5, 1971),46.

34. "Cornell Scientists See Thermal Pollution of Cayuga Lake by Planned Nuclear Power Plant". Conservationist, Volume 23 (August 1968), 2-5.

35. Costaglio, F. "Nuclear Power...The Information Communication Predicament". Address, February 17, 1971, Vital Speeches,Volume 37 (May 1, 1971), 436-41.

36. Coughlin, W.J. "Perils of Plato". Missiles and Rockets, Volume 14 (June 22, 1964), 46.

37. Curtis, Richard and Hogan, Elizabeth. "Myth of the Peaceful Atom". Natural History, Volume 78, No. 4 (March 1969), 6-8.

38. "Destruction on Jackass Flats; Nuclear Reactor Kiwi Blown Up". Time, Volume 85 (January 22, 1965), 35.

39. "Dig They Must ... Nuclear Plant in New York City". Business Week, (September 14, 1963), 82-4.

40. Dreher, C. "Atomic Power; The Fear and The Promise". Nation, Volume 198 (March 23, 1964), 289-93.

41. Eipper, A.W. et.al. "Impacts of Nuclear Power Plants on the Environment". Living Wilderness, Volume 34 (Autumn 1970), 5-12.

42. Eipper, A.W. "Pollution Problems, Resource Policy and the Scientist". Science, Volume 169 (July 3, 1970), 11-15.

43. "Fast Breeder Moving Fast Despite the Critics". Science News, Volume 100 (October 16, 1971), 71.

44. Feruis, L. "Bombs or Reactors?". Bulletin of the Atomic Scientist, Volume 26 (June 1970), 28-9.

45. "Four U.S. Nuclear Reactors Placed Under IAEA Safeguards". Department of State Bulletin, Volume 51 (July 6, 1964), 27.

46. "France Buries Its Pride". Time, Volume 94 (November 28, 1969), 96.

47. "From the H-Bomb: Power Without Pollution". Business Week, Volume 34 (September 12, 1970), 80.

48. Fusch, B.H. "A Bomb on New York's Doorstep; Ravenswood Power Plant". Science Digest, Volume 54 (October 1963), 29-35.

49. Gallagher, P. "Nuclear Power for Piguads: Pigua, Ohio". American City, Volume 79 (July 1964), 71-2.

50. Gannon, R. "What Are The Dangers?". American City, Volume 71 (January 1972), 36-42.

51. "GE's Nuclear Grant Warms Up; Oyster Creek Plant, Tom's River, New Jersey". Business Week, (May 17, 1969), 46.

52. Gendlin, F. "Palisades Protest: A Pattern of Citizen Intervention". Bulletin of the Atomic Scientist, Volume 27 (November 1971),53-56.

53. Gillette, R. "Nuclear Power in the USSR; American Visitors Find Surprises". Science, Volume 173 (September 10, 1971), 1003-6.

54. Gillette, R. "Nuclear Reactor Safety: A New Dilemma for the AEC". *Science*, Volume 173 (July 9, 1971), 126-30.

55. "Glutton for Punishment; GPH's Oyster Creek Plant". *Forbes*, Volume 105 (February 15, 1970), 34.

56. Gofman, John William and Tamplin, A.R. "Harrassment Charges Against AEC, Livermore". *Science*, Volume 196 (August 28, 1970), 838-43.

57. ______________. "Radiation: The Invisible Casualties". *Environment*, Volume 12, No. 12 (April, 1970), 19ff.

58. "Good and Bad Subsidies; Privately Owned Public Service Company of Colorado". *New Republic*, Volume 153 (December 25, 1965), 7.

59. Gough, W.C. and Eastland, B.J. "Prospects of Fusion Power". *Scientific American*, Volume 224 (February 1971), 50-64.

60. Graham, F., Jr. "Tempest in a Nuclear Teapot; Thermal Pollution". *Audobon*, Volume 72 (March 1970), 12-19.

61. "Great Breeder Dispute". *Time*, Volume 98 (November 1, 1971), 102.

62. Green, H.P. "Reasonable Assurance of No Undue Risk; Reactor Licensing Procedures". *Scientist and Citizen*, Volume 10 (June 1968), 128-40.

63. Gustafson, P.F. "Nuclear Power and Thermal Pollution: Zion, Illinois". *Bulletin of the Atomic Scientist*, Volume 26 (March 1970), 17-23.

64. Hanlon, J.O. "Atomic Bomb in the Land of Coal; TVA's Contract for the Biggest Nuclear Power Plant". *Fortune*, Volume 74 (September 1966), 132-3.

65. "Heating Up Lake Cayuga; Report on Cornell Pamphlet". *Scientist and Citizen*, Volume 10 (June 1968), 141.

66. Hedgpeth, J.W. "Bodega Head, A Partisan View". *Bulletin of the Atomic Scientist*, Volume 21 (March 1965), 2-7; Discussion,Volume 21 (October 1965), 27-8; and Volume 23 (December 1965).

67. Hogerton, J.F. "Arrival of Nuclear Power". *Scientific American*, Volume 218 (February 1968), 21-31.

68. Holcomb, R.W. "Power Generation: The Next 30 Years". *Science*, Volume 167 (January 9, 1970), 159-60.

69. Holden, W.M. "Hot Water; Menace and Resource". *Science News*, Volume 94 (August 17, 1968), 164-6.

70. Hunt, R.P. "Atomic Question for the City; Proposed Nuclear Power Plant in Queens". *New York Times Magazine*, October 6, 1963, 46ff.

71. Jacobsen, S. "Land Use Dispute in Illinois; Nuclear Power Vs. Crops". Bulletin of the Atomic Scientist, Volume 29 (January 1973) 43-5.

72. Jordan, W.H. "Nuclear Energy: Benefits Versus Risks". Physics Today, Volume 23 (May 1970), 32-8.

73. Klauster, A.P. "Radiation and Social Ethics: Conference of Nuclear Scientists and Theologians". Christian Century, Volume 80 (February 13, 1963), 199-200.

74. Kobler, J. "Four Deadly Minutes; Accident at the Boris Kidrick Institute of Nuclear Sciences, Yugoslavia". Saturday Evening Post, Volume 238 (November 6, 1965), 42-44.

75. Lambert, J. "Power and Politics; German Nuclear Research". Science News, Volume 94 (November 2, 1968), 454.

76. Langewiesche, W. "Can Our Rivers Stand The Heat?". Reader's Digest, Volume 96 (April 1970), 76-80.

77. Lapp, R.E. "Four Big Fears About Nuclear Power". New York Times Magazine, February 7, 1971, 16-17; Discussion, March 21, 1971, 79-80.

78. ________. "The Nuclear Power Controversy". New Republic, Volume 164 (January 23, 1971), 18-21;(February 6, 1971), 20-23; (February 27, 1971), 17-21.

79. ________. "Nuclear Power for Electricity". Current, Volume 127 (March 1971), 42-9.

80. Lawrence, D. "International Suicide?" U.S. News and World Report, Volume 63 (October 2, 1967), 116.

81. "Living With Atomic Power and Liking It; Yankee Atomic Nuclear Power Station; Rowe, Mass.". U.S. News and World Report, Volume 70 (March 29, 1971), 67-9.

82. Lubkin, G.B. "Columbia Reactor Gets Thumbs Down". Physics Today, Volume 24 (June 1971), 61-62.

83. Marine, G. "Outrage on Bodega Head". Nation, Volume 196 (June 22, 1963), 524-7; Discussion, Volume 197 (August 10, 1963), 62 .

84. Meier, R.L. "Social Impact of a Nuplex; Advantages of Nuclear-Powered Industrial Complexes". Bulletin of the Atomic Scientist, Volume 25 (March 1969), 16-21.

85. Nelson, B. "Thermal Pollution: Senator Muskie Tells AEC to Cool It". Science, Volume 158 (November 10, 1967), 755-6.

86. "New River; Sixteen Nuclear Power Plants on the Great Lakes". Environment, Volume 12 (January 1970), 36-40.

87. "No Rush to Nuclear Plants". Business Week,(February 8, 1964), 100+.

88. Novick, Sheldon. "Earthquakes and Nuclear Power". Scientist and Citizens, Volume 10 (November 1968), 230-41.

89. ____________. "Miles From Times Square". Adaptation of address of December 2, 1968, Environment, Volume 11 (January 1969),10-15.

90. "Nuclear Fueled Desalting Power Plant Planned". Science News, Volume 90 (August 27, 1966), 130.

100. "Nuclear Plant Builders See A $1 Billion Market". Business Week, (March 5, 1966), 144-6.

101. "Nuclear Power A.D. 2000". Science News Letter, Volume 86 (September 5, 1964), 149.

102. "Nuclear Power Loses a Battle in Court". Business Week, (July 31, 1971), 24.

103. "Nuclear Power: Setback? Comeback of Coal Plants". Forbes, Volume 102 (November 15, 1968), 55.

104. "Overselling a Dream". Science News, Volume 96 (August 9, 1969), 113-14.

105. "Patient Earth". Excerpts edited by T. Harte and R. Socolow, Physics Today, Volume 23 (December 1970), 28-9.

106. "Peaceful Atom, Friend or Foe?". Time, Volume 95 (January 19, 1970), 42-3.

107. Pesonen, D.E. "Atomic Insurance: The Ticklish Statistics". Nation, Volume 201 (October 18, 1965), 242-5.

108. "Power Reactor Makers Launch Business Drive". Business Week, (March 14, 1964), 43-4.

109. Pursglove, S.D. "A-Power: Creeping Giant". Popular Mechanics, Volume 119 (June 1963), 100-4.

110. Ramey, J.T. "Competitive Vigor in Nuclear Power; Unique Partnership of Business and Government". Harvard Business Review, Volume 46 (July 1968), 126-42.

111. Roe, K.S. "Bodega: Symbol of a National Crisis". American Forum, Volume 69 (December 1963), 22-5.

112. Schlesinger, J.R. "AEC to Referee, Not Promote Industry". Science, Volume 174 (October 29, 1971), 478.

113. Schubert, A.E. "What Nuclear Power Will Do To And For The Urban Environment". American City, Volume 83 (December 1968), 67-9.

114. "Scientists Question Reactor Effects; Study of Monticello Nuclear Generating Plant in Minnesota". Scientist and Citizen, Volume 10 (August 1968), 154-7.

115. Seaborg, G.T. "Nuclear Power Plants, Boon or Blight?". Natural Wildlife, Volume 9 (April 1971), 21-23.

116. Seaborg, G.T. "On Misunderstanding The Atom". Adaptation of address of March 22, 1971, Bulletin of the Atomic Scientist, Volume 27 (September 1971), 46-53.

117. "Sluggish Atom; Delays in Construction". Newsweek, (June 23, 1969), 81-2.

118. Smith, M.M. "Labor and the Savannah River AEC Project". Monthly Labor Review, Volume 88 (July 1965), 799-800.

119. Sporu, P. "Energy for Man and Environmental Protection". Science, Volume 166 (December 19, 1969), 1459-60; Volume 167 (March 13, 1970),1439.

120. Stearnes, B. "Heat Waste; Study of Turkey Point Plant's Effect on Biscayne Bay". Sea Front, Volume 16 (May 1970), 154-63.

121. Sternglass, E.J. "Health Effects From Radiation". Address of July 14, 1971, Vital Speeches, Volume 37 (September 1, 1971) 699-703.

122. ______________. "The Death of All the Children". Esquire, Volume 72 (September 1969), 1a-1d.

123. "Standardization Will End Nuclear Energy Growth". American City, Volume 84 (July 1969), 75.

124. "Switching to the Atom; Nationwide Surge". Time, Volume 88 (October 14, 1966), 100.

125. "Teller Attacks Breeder Safety; Nuclear Power Reactors, Sodium-Cooled". Science News, Volume 92 (August 26, 1967), 200ff.

126. Thomsen, D.E. "Fusion Power to Fight the Trash Glut; Self-Sustaining Fashion Reactor". Science News, Volume 97 (March 7, 1970), 249-50.

127. Tilden, P.M. "The California Atom Controversy: Bodega Head, Sonoma County". Natural History, Volume 72 (December 1963), 64-5.

128. "TVA Chooses Atomic Power". Business Week, (June 25, 1966), 150.

129. Wallace, W.A. "Radiation and Social Ethics: Conference of Nuclear Scientists and Theologians". America, Volume 108 (June 22, 1963).

130. Walsh, J. " Vermont: A Power Deficit Raises Pressure for New Plants". Science, Volume 173 (September 17, 1971), 1110 +.

131. "War Over the Peaceful Atom; Radioactive and Thermal Hazards". Life,(September 12, 1969), 2613-33.

132. Weinberg, A.M. and Hammond, R.P. "Global Effects of Increased Use of Energy". Bulletin of the Atomic Scientist, Volume 29 (February 1973), 4.

133. Wetboy, A. "Nuclear Power on Salmon Rivers". Nation, Volume 205 (October 9, 1967), 337-9.

134. Wheeler, H. and Burdick, E. "Politics of Destruction". Saturday Review, Volume 46 (November 9, 1963), 25-7.

135. "Who Gets the Last Radioactive Word? Minnesota VS AEC". Business Week, (May 31, 1969), 28.

Books and Reports:

136. Curtis, Richard and Hogan, Elizabeth. Perils of the Peaceful Atom. Garden City: Doubleday and Company, 1969.

137. Foreman, Harry, ed. Nuclear Power and the Public. Minneapolis: University of Minnesota Press, 1970.

138. Gofman, John W. and Tamplin, Arthur. Poisoned Power. Emmanus, Pennsylvania: Rodale Press, 1971.

139. ______________________. 'Population Control' Through Nuclear Pollution. Chicago: Nelson-Hall, 1970.

140. Holdren, John and Herrara, Philip. Energy. San Francisco: Sierra Club, 1971.

141. Independent Citizens' Research Foundation for the Study of Degenerative Diseases, Inc. Report of the Foundation, Danger from Atomic Power Plants, March 1969.

142. Metzger, H. Peter. The Atomic Establishment. New York: Simon and Schuster, 1972.

143. Lewis, Richard. The Nuclear Power Rebellion: Citizens vs. the Atomic Industrial Establishment. New York: Viking Press, 1972.

144. Nelkin, Dorothy. Nuclear Power and Its Critics: The Cayuga Lake Controversy. Ithaca, New York: Cornell University Press. 1971.

145. Novick, Sheldon. The Careless Atom. New York: Houghton-Mifflin, 1969.

146. Orlans, Harold. Contracting for Atoms. Washington: The Brookings Institution, 1967.

INDEX